What Every Engineer Should Know About

ENGINEERING WORKSTATIONS

WHAT EVERY ENGINEER SHOULD KNOW

A Series

Editor

William H. Middendorf

Department of Electrical and Computer Engineering
University of Cincinnati
Cincinnati, Ohio

Vol. 1 What Every Engineer Should Know About Patents, *William G. Konold, Bruce Tittel, Donald F. Frei, and David S. Stallard*

Vol. 2 What Every Engineer Should Know About Product Liability, *James F. Thorpe and William H. Middendorf*

Vol. 3 What Every Engineer Should Know About Microcomputers: Hardware/Software Design: A Step-by-Step Example, *William S. Bennett and Carl F. Evert, Jr.*

Vol. 4 What Every Engineer Should Know About Economic Decision Analysis, *Dean S. Shupe*

Vol. 5 What Every Engineer Should Know About Human Resources Management, *Desmond D. Martin and Richard L. Shell*

Vol. 6 What Every Engineer Should Know About Manufacturing Cost Estimating, *Eric M. Malstrom*

Vol. 7 What Every Engineer Should Know About Inventing, *William H. Middendorf*

Vol. 8 What Every Engineer Should Know About Technology Transfer and Innovation, *Louis N. Mogavero and Robert S. Shane*

Vol. 9 What Every Engineer Should Know About Project Management, *Arnold M. Ruskin and W. Eugene Estes*

Vol. 10 What Every Engineer Should Know About Computer-Aided Design and Computer-Aided Manufacturing, *John K. Krouse*

Vol. 11 What Every Engineer Should Know About Robots, *Maurice I. Zeldman*

Vol. 12 What Every Engineer Should Know About Microcomputer Systems Design and Debugging, *Bill Wray and Bill Crawford*

Vol. 13 What Every Engineer Should Know About Engineering Information Resources, *Margaret T. Schenk and James K. Webster*

Vol. 14 What Every Engineer Should Know About Microcomputer Program Design, *Keith R. Wehmeyer*

Vol. 15 What Every Engineer Should Know About Computer Modeling and Simulation, *Don M. Ingels*

Vol. 16 What Every Engineer Should Know About Engineering Workstations, *Justin E. Harlow III*

Other volumes in preparation

What Every Engineer Should Know About

ENGINEERING WORKSTATIONS

JUSTIN E. HARLOW III
Fairchild Camera & Instrument Corporation
South Portland, Maine

MARCEL DEKKER, INC. New York and Basel

Library of Congress Cataloging-in-Publication Data

Harlow, Justin E., [date]
What every engineer should know about engineering workstations.

(What every engineer should know ; v. 16)
Includes bibliography

Includes bibliographies and index.
1. Engineering--Data processing. 2. CAD/CAM systems.
I. Title. II. Series.
TA345.H318 1985 620'.0028'5 85-29257
ISBN 0-8247-7509-0

MARCEL DEKKER, INC.
270 Madison Avenue, New York, New York 10016

Current printing (last digit):
10 9 8 7 6 5 4 3 2 1

PRINTED IN THE UNITED STATES OF AMERICA

PREFACE

Computer-aided design (CAD) and computer-aided manufacturing (CAM) are terms that, by now, are familiar to most engineers and managers. Computer-aided engineering (CAE), however, is a new term which has appeared only in the last several years. Like most new ideas, and particularly new ideas in the fast-paced engineering field, CAE means different things to different people. Whether you think of CAE as merely a new, compact acronym to replace the cumbersome "CAD/CAM," or as a whole new technology, you will soon come to realize that the distinguishing feature of CAE in most people's minds is a new breed of equipment: engineering workstations. What are these systems that are attracting so much attention, and what do they offer to the engineering community that makes them so exciting?

There is no shortage of literature on the subject today; indeed, every engineering publication you pick up seems to have an article on workstations. Usually these articles present much detail and little insight. Nowhere is there an introduction to the subject that the busy engineer or manager can use to gain needed perspective. They need to understand this technology if their companies are to be competitive, yet they lack the time to read everything being written. This book is intended to help you understand, with only a minimum of effort, the important facts about engineering workstations and how they affect you.

My goal is not to make you an expert on engineering workstations; it is not to teach you how to use them, nor to tell you which system is best for you. Rather, I will survey in general terms the types of hardware and software that characterize engineering workstations, and which differentiate them from traditional CAD/CAM systems. I will describe applications for which engineering workstations provide appropriate solutions to engineering problems, and identify some for which engineering workstations may well be the wrong answer. I will briefly survey the types of tools provided the engineer by an engineering workstation, while avoiding the laborious detail that obscures the important concepts in so many engineering books today. I will explain some of the techniques used to justify the purchase of the equipment. Finally, I will describe the important trends in this volatile industry, to help you understand how it all fits together and where it is heading.

Today, as I prepare this manuscript for submission to my long-suffering publisher, the field of computer-aided engineering continues to profoundly change on a daily basis. Just as each chapter is completed, another newsletter or magazine seems to arrive, heralding yet another amazing product announcement. By the time you read this, much of the information I have included on specific products may well be dated. My advice to the reader, then, is to look beyond the details for the fundamentals I offer here. Look at the trends, the applications, the ground that has already been covered as you try to understand the present and the future.

Many readers will no doubt feel that I have focused too heavily on electronics applications for engineering workstations, and have ignored their particular specialty. What you observe is not tunnel vision, but a consequence of my attempt to write a timely book about a fast-changing subject. To date, nearly all the full-function engineering workstations on the market focus on electronics applications. But, as I write these words, there is a surge of new products for mechanical engineers, architects, and others that will change the entire face of those disciplines, as surely as the electronics industry is now being transformed. By reading about the impact these new tools have had on one industry, I hope you will infer the coming changes in your own.

This, then, is a survey book for managers and for design engineers, a guidebook for the uninitiated, a starting point. I hope you will find it easy to read yet informative, comprehensive yet unburdened by detail. Above all, I hope that you may learn enough to make intelligent decisions related to this new technology for the benefit of your company.

Justin E. Harlow III

CONTENTS

ABOUT THE AUTHOR

Justin E. Harlow III is Manager of CAD Engineering at Fairchild Camera & Instrument Corporation in South Portland, Maine, where he is responsible for the network of engineering workstations for electronic design and the CAD facilities for electronic, mechanical, and facilities design. Previously, he worked for Harris Corporation and ITT Semiconductors, and held engineering positions involving design, product engineering, test development, and manufacturing pilot line development. He is a member of the Institute of Electrical and Electronics Engineers, American Management Association, and honorary Computer Graphics Pioneers. Mr. Harlow received the B.E.E. degree (1964) from the University of Florida and M.B.A. degree (1983) from Florida Institute of Technology.

What Every Engineer Should Know About

ENGINEERING WORKSTATIONS

1

INTRODUCTION

Computer-aided design (CAD) and computer-aided manufacturing (CAM), the subject of another book in this series [1], has become a necessity in many industries. The electronics industry, in particular, requires computer automation in nearly every phase of design and manufacturing today. Electronics, more than any other industry, has assimilated its own products and used them to create their own successors. The computer and communications explosion that has occurred since the late 1960s has not only created the need for ever more powerful CAD and CAM tools, but has also provided the means to implement those tools.

Early CAD tools were batch programs which ran on mainframe computers. For instance, the pioneering ECAP program performed analog circuit simulation using the computational power of a large computer as a substitute for the trial and error methods of "breadboarding" a prototype circuit. Using ECAP, the engineer could build a simulated circuit using mathematical models of electronic components, and wire them together using a connection list. The information was communicated to the computer through a deck of punched cards, and the output was a textual listing of the voltage at each node and the current through each branch, for a series of time points. Using this information, the engineer could draw a graph illustrating the behavior of the model circuit under conditions he defined in the input deck.

Use of programs like ECAP enabled a small number of fortunate design engineers in large companies to explore optional circuit configurations, to experiment with component values, to determine behavior over a temperature range, and to optimize at least a few important circuit parameters in a way which would have been prohibitively time consuming and expensive using manual methods. However, to the great mass of engineers, large computers were rarely available or were far too expensive to be used routinely.

In the mid-1960s, two revolutionary advances in computing made the use of tools like ECAP possible for a large number of engineers: The development of timesharing on large computers, and the introduction of small scientific computers like the IBM 1130. The wide availability of these powerful, affordable computing resources brought rapid progress in the number and quality of programs for simulation and design. The term CAD was born in this era, just before the first 16-bit minicomputers appeared in 1968 and fueled the next revolution in CAD.

The appearance of compact, affordable computers like the Digital Equipment Corporation PDP-11 and the Data General NOVA made possible a revolution in CAD which continues to this day. Now it was possible for even companies of modest size to buy a minicomputer capable of running simulations in-house. Entrepreneurial companies offered self-contained graphics systems based on minicomputers, allowing many companies to mechanize much of their drafting and to produce their own printed circuit board artwork. These systems were offered both as generalized graphics systems, which could be used to put lines, curves, and text together in any form desired, and in specialized form as mechanical design systems, plant layout systems, schematic drafting systems, and so on. Soon, the potential of minicomputer-based drafting systems was extended to include three-dimensional drafting and some CAM applications such as automatic generation of numerical-control (NC) drilling and machining tapes.

Throughout the 1970s, CAD/CAM companies like Applicon Inc., Calma Co., Computervision Corporation, and many others continued to provide a steady stream of enhancements to their systems. By the end of the decade, most of these systems had gone far beyond their original mission of mechanizing the drafting process and were capable of being used for highly sophisticated design activities. In some companies, engineers used the systems for conceptualization and initial sketching of complex designs. This proved particularly beneficial in three-dimensional design, since the manual processes of projection and multiple-view generation could be totally automated by the computer. Another application in which CAD systems were heavily used by design engineers was the physical layout of electronic circuits, particularly integrated circuits (IC). Designing these tiny but complex parts was becoming a task too large for any sort of manual technique, and IC designers embraced CAD enthusiastically. The power and sophistication of such readily available CAD accelerated the pace of development in computer companies and the resulting flow of new and better systems enabled the development of still better CAD and CAM systems.

Until the early 1980s, however, even the powerful minicomputers and CAD/CAM systems were a shared resource. Their popularity, and the demand for ever more thorough design, assured that there was never enough horsepower available for all users at all times. The development of personal computers in the late Seventies suggested what the future held. At least for undemanding computational jobs like word processing, personal computers could be dedicated to an individual; they were always available on demand, and were never degraded in performance by other users. Several groups of talented design engineers possessing an entrepreneurial spirit recognized that the characteristics, if not the computational power, of personal computers were highly desirable for engineering computing. Capitalizing on the development of powerful 16-bit and 32-bit microprocessors (Figure 1.1), they pioneered the development of personal computers for engineers, and thus, the subject of this book: engineering workstations.

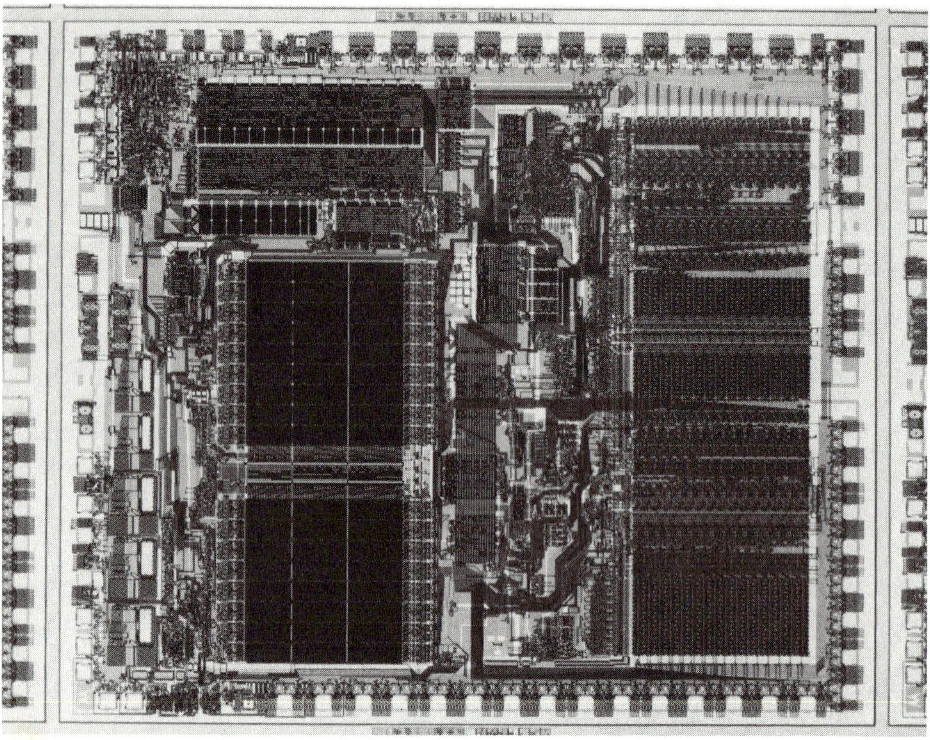

Figure 1.1 The Motorola MC68000, a single chip microprocessor with a 32-bit internal architecture, is the most popular microprocessor for engineering workstations. (Courtesy Motorola Inc.)

Concepts of CAD/CAM were developed in the 1960s, while the powerful hardware and software developed in the 1970s proved these concepts and demonstrated to the engineering community that design automation was a practical necessity. The tools have been developed and the hardware technology refined. Now, the CAD/CAM challenge for the 1980s is to make the tools developed over two decades readily, routinely, and exclusively available to any engineer who needs them.

1.1 Definition of an Engineering Workstation

An engineering workstation is an interactive graphics system intended for use by engineers in the design of products. Usually, but not always, these systems are based on powerful personal computers or compact minicomputers. In a sense, any turnkey CAD/CAM system of the type common in the 1970s could be thought of as a group of engineering workstations sharing a common host computer, but those systems are intended for a different type of user (Figure 1.2). While traditional CAD/CAM systems are used

Figure 1.2 Engineering workstations are typically based upon compact, powerful computers like these Apollo Computer Inc. products. (Courtesy Apollo Computer Inc.)

to increase the productivity of drafting and manufacturing departments, engineering workstations are intended to be used directly by engineers in the creative and conceptual phases of product development.

1.2 Computer-Aided Engineering

Workstations are associated with computer-aided engineering (CAE), a term commonly used to describe the use of computers to aid *creative* portions of engineering (as opposed to clerical functions like drafting). In this context, CAE includes simulation and front end verification tools, while CAD or CAD/CAM deals with physical layout, documentation, manufacturing interfaces, etc. Despite the somewhat arbitrary distinction though, the two disciplines are interrelated. Feedback from CAD/CAM (the back end activities) to CAE (the front end) is an exceptionally powerful concept.

For instance, an electronic engineer can use a circuit or logic simulator (or both) to model critical portions of a design prior to building hardware, using assumed values for the capacitances and inductances that will be present in a physical printed circuit board. Many CAD/CAM systems include software which derives electrical parameters from the lines and shapes used to layout circuits. An engineer uses these parameters to perform post-simulation on the original design. Often, the influence of physical parameters is sufficient to significantly change the performance or even the function of a circuit. Designers can now allow for such effects in a rigorous, quantitative fashion that was impractical only a few years ago.

Manufacturing engineers and mechanical designers using workstations in the near future will be able to develop tool paths and work flows during the design of components and assemblies. Simulation of tool movements at this stage of product development often uncovers interference between workpieces and fixtures, and allows engineers to correct manufacturing problems before a design is even completed.

The rapid proliferation of robots in factories has spawned an entire new CAD/CAM application area: the design of work cells. Engineers can now work with computer-simulated models of manufacturing equipment to develop optimum work cells and process flows. Some systems present startlingly realistic visual images of the robot in its work environment, even simulating its motion.

All of these CAE applications are fundamentally different from the drafting-oriented applications of traditional CAD/CAM. Engineers are the users of the tools, and the tools require great computer power. Such techniques potentially have enormous impact on engineering productivity, and on the quality and complexity of products that can be designed. Many companies are now offering exciting products to exploit this new kind of engineering: inexpensive, powerful engineering workstations.

1.3 Functions of an Engineering Workstation

To describe the functions of an engineering workstation is to describe what an engineer does for a living; an ideal workstation would provide all the "tools" used by the engineer, not only on a project, but in his routine daily activities.

Some studies have indicated that the average design engineer spends less than 25 percent of a typical day doing actual design and analysis work; the remainder of his day is spent on a variety of nontechnical chores such as writing reports, attending meetings, planning activities, and talking to coworkers. In engineering, as in other fields, the number of interactions required between team members determines the efficiency of their work: more than five or six interaction pairs can reduce team efficiency to the point that almost no productive work is accomplished. The communications and database management capabilities of a multi-workstation network permits the structuring and control of project communications, thus increasing overall team efficiency.

Because the engineer's time is spent in such a variety of activities, a workstation that will improve his effectiveness must address more than just the drafting and simulation tasks. The true significance of engineering workstations can only be appreciated when they are viewed not as collections of programs running on small computers, but as an environment and the embodiment of a philosophy. In a broad sense, then, the workstation environment provides for the engineer six fundamental sets of functions: specification, analysis, verification, communications, documentation, and partitioning.

1.3.1 Specification

Any function of an engineering workstation which involves the description or conversion of a design can be considered a "specification" tool. Specification tools include mechanical drawing editors, schematic editors, and symbol library editors, as well as bill-of-material processors and netlisters. Although electrical and mechanical drawings are produced using the editors, it is important to note that, in a CAE system, what appear to be simply drawings are actually graphic representations of a design specification; that is, much information regarding materials, properties, component values, part numbers, etc., are incorporated in order to provide information to analysis tools. Thus, unlike traditional CAD/CAM systems, CAE systems use drawing for much more than documentation.

The difference between specification drawings and documentation drawings can be best illustrated through examples. For instance, a draftsman detailing a part uses lines, curves, points, and text to produce a drawing; most CAD/CAM systems do exactly the same thing. A CAE system, however, will require the user to supply further information which describes what the part is, how it behaves, its properties, how to build it, and many other details. The draftsman might have lettered this information on his drawing; the CAE system stores such information in a database and indicates its presence on drawings with text. The appearance of the two drawings may well be identical, but the CAE system's analysis tools can utilize the stored data directly and automatically.

Advertisements and demonstrations of engineering workstations for electronics applications frequently dwell on the usefulness of the systems for "schematic capture," a term which has come to imply more than simply drawing circuits. Many full-function workstations evolved from systems whose original function was simply to give engineers an easy, graphical way to specify their design ideas to various software packages. The user of a schematic editor places symbols from a library at desired locations on a simulated drawing and connects them with lines. On a traditional CAD/CAM system, these would simply be lines and symbols; and their connections as represented by the lines. Modifying the drawing using the schematic editor also modifies the design in the system's data base. The drawing is the "window" through which designs are viewed and manipulated.

Figure 1.3 Analog Design Tools' Analog Workbench is used for analysis and design of electronic circuits. (Courtesy Analog Design Tools, Inc.)

1.3.2 Analysis

The most computer-intensive portions of an engineer's job, the analytical functions, were historically the first portions which were automated. For the electrical engineer, a workstation may provide circuit analysis (Figure 1.3), logic simulation, timing analysis, and fault simulation. Details of all these tools will be described in the chapter on electronic engineering workstation software.

Workstations providing analytical tools for the mechanical engineer were beginning to emerge in 1984. In addition to drafting, these new mechanical engineering workstations will have tools for calculating moments, areas, volumes, and masses. Some systems feature "kinematics," which are simulations of the motions of parts in an assembly.

A major impediment to the development of true mechanical engineering workstations, however, is that few economical, personal computers yet have the required local computational power to perform finite-element analysis of stresses and heat flow. While there are systems currently offered capable of analyzing small parts and assemblies, their capacity is still quite limited. These systems foreshadow the coming of engineering workstations that will have extensive analytical capabilities, but such systems will not be widespread until much more powerful small computers are available at a reasonable price. Until then, many of the analytical functions used by mechanical engineers will still be done on shared mainframe computers.

Several vendors now offer compact engineering workstations specifically for architects and industrial engineers. These systems incorporate analytical functions tailored to these applications, such as stacking and blocking and space planning. An architect may specify to his system information on departmental relationships within a proposed building, and determine optimal placement of offices and floors. Industrial engineers may utilize optimization tools to tune the flow of work on a factory floor.

1.3.3 Verification

Much is being written today in the engineering press about making products correct by design; however, no system can yet force designs to be correct, and virtually no design work is yet done without extensive checking for mistakes or incorrect assumptions. Thus one of the most important responsibilities of an engineer, and of an effective workstation, is design verification.

For instance, many electronic engineers use logic simulators to verify the correct function of their designs. The design engineer, having interconnected a set of components, composes a set of desired circuit inputs and a set of expected outputs. The logic simulator quickly models the behavior of the design and compares the results to the engineer's expected outputs. This use of a logic simulator is, of course, traditional and commonplace in many electronics companies. However, most engineering workstations provide a friendly environment (Figure 1.4) in which to perform the verification, capitalizing on close communication between tools. Some systems present results in a form resembling an oscilloscope display. The engineer can easily place simulated probes in the sche-

Figure 1.4 Many engineering workstations utilize a multi-window screen display to enhance communication between the engineer and the various tools he uses. (Courtesy Analog Design Tools Inc.)

matic drawing, much as he could do with a physical breadboard circuit, and see the waveforms at the selected points on the display.

Other tools for electronic design verification examine the physical implementation of a system in detail and compare it to a set of design rules. For instance, Design Rules Check (DRC) and Electrical Rules Check (ERC) tools examine layouts of integrated circuits and printed circuits for violations of wire-spacing rules, overlap and via rules, and the like. Electrical rules checked by ERC tools include correct construction of transistors in an integrated circuit, open and short circuits to power busses, and the like.

While mechanical engineering workstations are not yet widespread, many design verification tools have been developed for use on CAD/CAM systems and will continue to be used on the newer systems. For instance, a designer may create a set of parts which are intended to fit together in an assembly; frequently, it is impractical or even impossible to visualize the spatial relationship of the parts as the mechanism operates. So-called kinematics packages are used by engineers to verify that all parts of such an assembly fit together properly and do not interfere or jam when the mechanism is operated. Many CAD/CAM systems demonstrate the mechanism on their display screens, showing all parts fitting and moving together. Jams and other design problems are visually highlighted, allowing the engineer to redesign or modify the offending parts. Similarly, engineering workstations for architectural and plant engineers will include verification tools for piping rules and interferences, electrical code compliance, duct sizing, and many other potential sources of design errors.

1.3.4 Communications

In the broad sense, communications is the foundation upon which all other workstation functions rest. Communications in this context implies not only interchanges between people, but between machines and between programs (tools).

Most manufacturers of workstations stress "integration of CAE tools," or their "integrated database" as major selling points. For purposes of classification and understanding, these concepts should properly be regarded as communication activities since their real function is to allow easy, error-free communication between users, data files, systems, and software packages.

Netlisters, for example, are specialized routines to provide communication between electronic schematic databases and other programs. Netlisters extract information on circuit or logic elements and their connections from the database and format the data to drive other tools such as simulators, network verifiers, test programs, and layout programs. The converse of a netlister has no commonly accepted name, but is used to perform *back annotation* of the original database. For instance, a program may extract physical information which was created by a layout program, and feed it back to the schematic database for further simulation studies of the design.

Many communication functions of the typical engineering workstation are familiar applications found on ordinary personal computers. Electronic mail is becoming fairly common in office environments utilizing large computers and networks, and in engineering workstation networks as well. This technology allows messages to be composed using a text editor and sent to another individual or to a distribution list. A person receiving an electronic message is usually signalled on his terminal when the message arrives, and periodically thereafter until the message is handled. Receiving parties can read, forward, delete, or file their messages without ever dealing with paper. Most engineering workstations are capable of being interconnected with other workstations, with mainframes, or with personal computers; such interconnections may be local, using coaxial cables or simple wire pairs, or remote, using modems. This interconnection permits very elaborate electronic mail systems to be implemented, and, indeed, most workstations provide this valuable tool for communication among project team members and between engineers and managers.

Electronic mail and telecommunications provide the means for people involved in a project to work together regardless of location or time of day. Most workstations provide at least some capabilities in these areas, and many permit literally hundreds of designers and managers to communicate over extensive networks. While large computer systems have provided these features in some form for many years, the key difference with respect to the engineering workstation is that these capabilities are provided as an integral part of the engineer's tools, not as a separate environment. Thus, sending a report to corporate management across a continent or reading a message from a coworker in the next building no longer requires the use of a different computer or waiting for clerical personnel. In many cases, the engineer need not even stop what he is doing—for instance, running a simulation—to use the communications facilities of his workstation.

1.3.5 Documentation

We used a broad definition of communications for classification purposes. Similarly, documentation can be given a broad meaning. It can be said that documentation provides the link to the future and the past that communications provides with the present. In this context, we will include production of drawings and schematics, online "engineering notebooks," aids to writing (for technical and user manuals, specifications, and application notes), and database systems.

The word processor software which runs on a typical engineering workstation is much more elaborate than its counterpart on an office PC. In the chapter on software tools, elaborate documentation systems will be described, such as programs which permit sections of design schematics and other graphic images to be included in textual documents.

Database management is a crucial documentation function that a workstation environment can provide (although many do not do so). Some of the more elaborate and costly engineering workstation products maintain a unified database of all information for making up a project. Included in such a database are the graphics images for drawings and schematics, the connection lists, the simulation models, the design and manufacturing parameters, and the bill of materials. In addition, such a data base will contain a configuration specification, a "roadmap" which specifies how all the various pieces of information fit together, and which versions make up a particular release of the product.

An engineering database is very different from the neat, fixed structures studied in data processing classes. Engineering design data are by nature highly changeable. In general, no fixed schema can be written for design information; the information and the relationships among data are dynamic; there is no concept of a transaction in engineering data which corresponds to the small number of types of database updates usually defined for commercial data. For these reasons, some vendors avoid the database issue altogether, and maintain simple files for each type of application software; translators and converters are used to communicate among tools.

In an ideal system, building a design and redefining relationships among data would be a dynamic process. No conversions between software tools would be needed, since all data would be represented and stored in a consistent way. In fact, some systems currently available come very close to this ideal, and provide not only efficient design tools, but the ability to reconfigure the database and the system environment without changes to design databases in progress. Tools are provided for the management of database versions, design configuration, releases, and archiving.

An important documentation function of some engineering workstations is the "electronic journal," a combination of word processing, drafting, and, sometimes, electronic mail which can be used by the engineer to document his concepts and ideas much as he would do in a notebook. Such packages are ideally available to the user of the workstation at all times, regardless of whatever other activity he may be engaged in. To be useful, the notebook must be as accessible while entering a schematic as when drafting a letter. (In the parlance of software specialists, such a system is "context free.") Electronic journals, almost by definition, require that the workstation be capable of simultaneously supporting multiple, active processes; thus, they are found only on the more elaborate and expensive systems today.

1.3.6 Partitioning

Most of the engineering workstation tools today stress capture and simulation of designs. However, as the complexity of the systems being designed increases, the next major advances in engineering workstations must be tools that aid in the partitioning decisions crucial to large designs [2]. Partitioning problems involve such issues as tradeoffs between hardware and software implementations of a function, mechanical and packaging considerations which depend on the size and count of parts, and allocation of functions to particular subassemblies or circuit boards.

Most engineering workstations for electronic design encourage design partitioning through the representation of complex designs as a hierarchy. A hierarchical database can be visualized as a treelike structure, in which the leaves are the simplest components of a design, the stems are the connections which tie components to larger structures or branches, and the main trunk of the tree is the entire design. Viewing a design in this way, a designer can deal with large parts of a system (branches) in an abstract way while working out the overall arrangement of the system. He need not be concerned with the arrangement of individual components or gates (leaves) when he draws a register-level diagram of a computer; conversely, he may deal only with a small portion of circuitry to optimize a delay path without worrying about the entire system. This treelike data structure allows useful abstractions to be made in dealing with high complexity levels.

However, for hierarchical partitioning of systems to be a valid and useful tool, conformance with top-level specifications and partitions must be maintained at the bottom (physical) level. For instance, it does little good to partition a system carefully into subassemblies if it is possible to later make a change to a subassembly which will cause it not to fit in the overall system. Future engineering workstations will address such issues of consistency automatically, at least for disciplines in which the hierarchical design style is useful, such as electronic systems.

Tools supporting decisions made at the top of a hierarchy similarly must be capable of accurately assessing consequences at the bottom (size, power, pinout, etc) and avoid actions by the engineer which invalidate the overall design constraints. For instance, changes in circuit connections made at the top level of the hierarchy must not be permitted if, as a result, lower-level components already in use cannot maintain the target system performance after the change.

Automatic enforcement of partitioning decisions throughout a system is not yet available, although various tools are emerging which deal with partitioning and abstraction through portions of an electronic design hierarchy. For instance, some logic simulators allow use of functional models (abstract descriptions of what a block is supposed to do, as opposed to how it is built) intermixed with individual logic gates. In effect, such simulators are dealing simultaneously with different levels of a hierarchy. Physical modeling systems, (hardware attachments to an engineering workstation that allow physical components to be used in place of models in a simulation), allow designers to evaluate system performance at a very high level without dealing with the low-level details of complex subsystems such as microprocessors.

Partitioning, however, should not be limited only to electrical subdivisions of a system. The electronic component costs in a typical system today can be forced down easier than the cost of the mechanical components; thus, it is crucial that a really complete design system account for mechanical as well as electrical factors throughout the design cycle. All manufacturing and implementation costs must be considered, not just the semiconductor component design costs. Future engineering workstations, perhaps a decade from now, will feature tools for both mechanical design and electrical/electronic design which work from a common database and permit system designers optimal capability to partition their designs.

References

1. J. K. Krouse, *What Every Engineer Should Know About Computer-Aided Design and Computer-Aided Manufacturing*, Marcel Dekker Inc., New York, 1982.
2. A. S. Rappaport, Computer-Aided Engineering Tools, *EDN Magazine*, September 15, 1983, p. 120.

2

REASONS FOR (AND AGAINST) ENGINEERING WORKSTATIONS

After covering briefly some of the things that engineering workstations can do for the engineer, it is easier to understand exactly why it is that such systems have become so enormously popular in such a short time. This brief chapter covers some of the key factors managers must consider before making a decision to acquire engineering workstations for their organizations. We will look into some of the reasons companies choose to use workstations, and some important issues regarding the impact of workstations on people and organizations. Finally, some cautionary information is presented about the vendors who offer the equipment.

2.1 Complexity Management

The design of large systems is a challenge not only to the technical expertise of designers, but to the skills of project leaders in managing the interrelations between subprojects. Historically, the problems of complex-systems development have been dealt with by large teams of people, and have been associated with projects such as construction, weapons systems, and shipbuilding. Although many of the same project management techniques developed for these huge systems have been applied successfully for the development of large electronic systems, it is only recently that the complexity of physically small and relatively inexpensive systems has become so high that a similar approach to their development has been needed.

The fundamental principle of complexity management is divide and conquer, that is, subdivide the large and intractable into the smaller and more comprehensible: huge projects become a series of subprojects which can be easily understood and dealt with. On very large projects, subdivision into comparatively small units can occur multiple times. As a simple example, the development of a spacecraft system could be divided into the ground support systems, the craft itself, and the recovery systems. Each of these is still an enormous project and would be further subdivided; the ground systems can be divided into earthworks, buildings, roadways, utilities, etc. At this point, the project manager in our simple example might hire several subcontractors for the partitions of the ground systems project, and then deal with each project as a whole. The manager's job now has become one of coordination, making sure the pieces fit together and will be completed in the right sequence and at the right time.

Project management requires highly-structured communication between project team members, as well as communication up and down a hierarchy of teams. Project components designed at one level must accurately fit the overall framework of the project. As we observed in Chapter 1, many of the capabilities of engineering workstations are ideally suited to just these types of communications. While a huge construction project spans many years and employs hundreds or thousands of people, the development of an electronic system usually is a fast-paced process done by small teams, or even by individuals. A network of engineering workstations has become a very attractive way to allow a design team to work on subdivisions of a large project in a controlled fashion.

Ideally, all data for a design task are maintained in a common database, and the design system provides a means for specifying and enforcing interface standards. Real engineering workstations, while falling far short of the ideal, provide useful tools for project management. Typically, a hierarchy of component libraries can be built and shared throughout a network (or across multiple users on a large, centralized system). Access to these libraries can be controlled in a variety of ways, but as a minimum it is possible to limit the privilege of modifying a library's contents to a "librarian." In many cases, even in the absence of a software system for controlling revisions, a manual bookkeeping system and a privileged librarian can control the overall configuration of a project.

The current trend in engineering workstations is toward a physically distributed database; that is, each station has a private mass storage device and file system. These distributed databases are almost universally organized into a hierarchy of directories (analogous to the branches of a tree) and files (analogous to the leaves). In these structures, there is essentially no distinction between a directory and a file, nor between a directory and a physical device. Thus, even though the system consists, in fact, of a number of linked smaller systems, the data structure can be made to behave exactly as if it were a single database on a large machine. Such file systems lend themselves very naturally to complexity management.

Continuing the simple example of the spacecraft project, the project manager might set up a top-level project directory and four subdirectories for the major subprojects. The spacecraft subdirectory might contain several more subdirectories, which contain yet more subdirectories, and so on. Eventually, this hierarchy reaches a point at which individual engineers are dealing with distinct projects. At this point, an engineer might have on his workstation a directory for, say, a particular control system. This engineer might maintain a directory for the released version of his controller, and a copy of the directory for changes in progress for a new version. The project manager, who is responsible for the overall systems project, must have a means to assure that, when he releases specifications for construction or manufacturing, he has the right version of every subsystem. A software system available to project leaders on some engineering workstation networks permits control of configuration, a roadmap which details exactly which files in which subdirectories, and which versions of them, are used to make up a particular version of the overall project.

Although today's engineering workstation networks can neither manage projects as large as the example used here, nor provide, in general, a complete set of tools for complexity management, their structure and philosophy are well suited for the development of such tools. In the next few years, users can expect to see engineering workstation vendors expend a great deal of effort in a race to develop sophisticated project management software.

2.2 Semicustom Integrated Circuits

The rapid pace of technological change in recent years has forced companies to compress the development cycle of increasingly complex products into shorter and shorter time spans. The manufacturer who offers a new type of product before his competitors has the opportunity to build market share quickly, a crucial factor in markets where life cycles are only a few years. The electronics industry provides an illustration of a technology that addresses the problem of rapid system development, a technology largely made possible by the widespread use of engineering workstations.

Semicustom integrated circuits are components of great complexity built to satisfy the unique system needs of individual customers, yet they can be made in large volume by semiconductor manufacturers. The semiconductor manufacturer gains many of the economies of scale normally associated with standard products, while the end user has the advantage of rapidly building specialized functions into new systems. These seemingly contradictory advantages for both parties are the strength of a class of products called gate arrays.

Gate arrays are integrated circuits which consist of a matrix of simple functional blocks. A particular function desired by a system builder is implemented in a gate array by adding one or more layers of interconnection atop a previously fabricated array.

Since all arrays of given a family have the same primitive structure, the semiconductor manufacturer is able to build and inventory a large number of identical wafers; these wafers are then customized quickly for individual customers by adding the final two or three masks.

Engineering workstations have fueled the development of gate array technology in recent years. The close coupling between design data and layout data, the communications capabilities, and the high availability of the workstation make it ideally suited for the design of gate arrays. Many semiconductor manufacturers now provide libraries of models and symbols for their gate array products, suitably formatted to run on popular brands of engineering workstations. Using these tools, the system designer can use arrays produced by a variety of manufacturers without mastering the technological details of any.

Until recently, the design of integrated circuits was a rare and restricted skill. Only the very largest systems companies were able to customize circuits for their products. There are said to be a hundred times as many accomplished system designers as there are expert chip designers. Gate arrays and inexpensive engineering workstations have changed the economics of the electronic systems business radically in only a few years.

The ability of system designers to specify and create special functions in silicon for a product also results in better systems. The designer using these tools need not be limited by the circuit functions a semiconductor manufacturer offers as standard parts, nor need he incur the enormous expense and risk of custom circuit development.

Although engineering workstations are acknowledged by most engineers to be beneficial to system design, the manager contemplating the use of these systems should be aware that not all engineers may be equipped to cope with all the complexities of semicustom technology. This awareness is an important factor in deciding whether your company is ready for engineering workstations.

2.3 Small Firms

Many small architectural firms, surveying offices, and contractors have been unable in the past to exploit CAD/CAM techniques simply because of the cost. Even a single terminal drafting automation system, until very recently, was a $100,000 investment. Such systems were not only costly to buy, but were physically large and required substantial facilities support (power, air conditioning, sound isolation, etc). Because of the cost of owning CAD/CAM systems, small engineering firms have been at a decided disadvantage relative to larger competitors in preparing bids and in producing design documentation.

Engineering workstations, particularly those based on ordinary personal computers, are ideal for the small firm. They are inexpensive to buy, rarely require special facilities, and are much easier to use than their ancestors. In fact, many small firms find that their existing personal computers can be upgraded to excellent drafting systems for only a few thousand dollars. Systems like AutoCAD™, a product of Autodesk Inc., have been accepted enthusiastically by large and small firms alike. Many large firms are even finding that the half-million dollar systems they once used can be replaced by ten-thousand dollar personal computers.

2.4 Human and Organizational Impact

It may seem obvious that design engineers are the intended users of most workstations. However, in many companies there are factors that dictate the use of workstations by other people. Deciding to use engineering workstations in a given company, and deciding how to use them, is heavily dependent upon the training and education, the work habits, and the desires of the intended users. It also depends heavily on the traditional organization and division of duties within the company.

Many companies begin with the assumption that engineers will be the exclusive users of engineering workstations, and that the scope of their work will broaden. This implied choice usually dictates that all workstations bought must have all the capabilities an engineer is likely to need to do a complex design single-handedly. Initial purchases of electronic workstations thus typically include schematic processing, simulation, physical layout, verification, communications, documentation, and so on. Mechanical and other types of workstations are not yet as complete as those offered for electronics, but the lessons of the CAD/CAM industry indicate that a similar pattern will emerge when the systems are available: first-time customers will order not only drafting, but a full line of finite element analysis, solid modeling, shaded image generation, NC tool programming, group technology, etc. In summary, first-time purchasers tend to buy a few of the most expensive workstations available from their chosen vendor; because of the cost of the systems, they then usually attempt to share them among many designers. This paradoxical situation, sharing systems which were designed to be dedicated to individuals, has hindered the widespread use of workstations in many companies.

Few companies have the means or the will to buy more than a few stations initially. Soon after the systems are in place, these companies find that it is difficult, if not impossible, for an engineer to do all aspects of a design on a workstation which is not dedicated to his use. Furthermore, it is also readily apparent that a single designer cannot simultaneously make use of all the tools and power of a full-function workstation. Because of these factors, many companies soon buy additional workstations which have substantially lower functionality than their first systems.

Limited-function workstations, linked through a high-speed communications network to more powerful computers, offer an attractive alternative to top of the line workstations. Using limited-function workstations for schematic editing or drafting, along with more powerful "servers" for simulation and other computing-intensive tasks, permits the use of different people for different tasks. It is common in many companies to use technicians or drafting specialists to enter and detail drawings, relieving engineers of those tedious tasks. In other organizations, product planners and marketing engineers may use inexpensive workstations to do conceptual work at the block-diagram level prior to beginning detailed design by design engineers. Clearly, these are good applications for low-cost workstations such as those based on personal computers.

Some companies choose personal computer based systems for their design engineers as well, not necessarily because they can perform all the required tasks, but simply because they are cheap enough that every engineer can have one [1]. Others feel that the cost of even a full-function workstation is minor when compared to the cost of recruiting and compensating a good design engineer (Figure 2.1); thus, these companies may choose to provide each

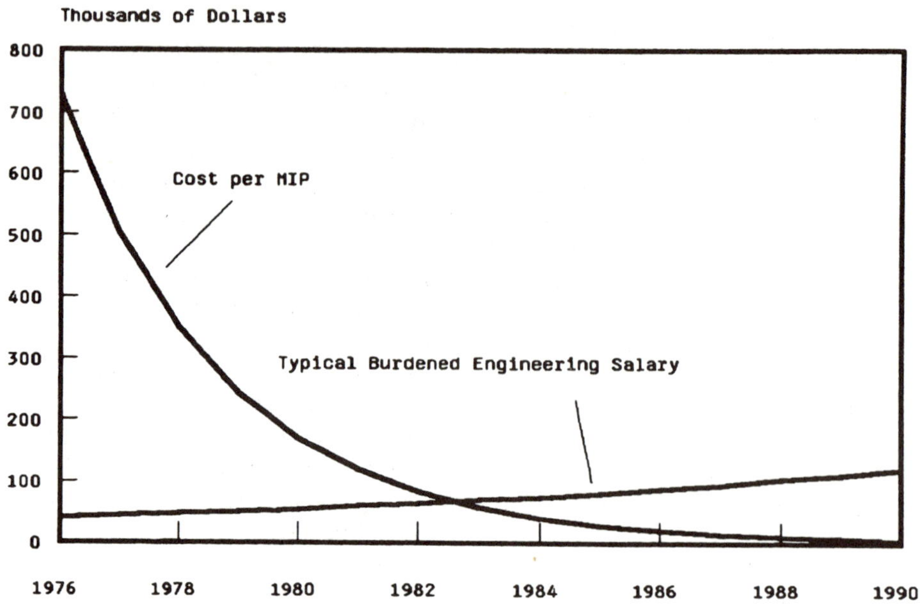

Figure 2.1 The cost of compensating a design engineer frequently exceeds the cost of today's CAE workstations. Engineering productivity improvements resulting from the use of engineering workstations can often pay for the equipment in less than a year. (Source: Dataquest Inc.)

engineer with a powerful workstation, capable of performing all the tasks he needs.

There is a tendency among many managers and educators to assume that engineers, given a complete and well-integrated set of design tools, can naturally deal with vastly broader problems than they could previously have handled. It is fashionable, particularly in electronics, to talk about "tall thin designers," renaissance men who deal with the entire hierarchy of a system design from the design of chips all the way to the packaging and cooling of equipment racks. However, most experts in the engineering workstation industry are considerably less sanguine about their products than these theorists.

Experienced system designers are not necessarily good detailers or chip designers; giving them a vast array of computerized tools usually will not change these fundamental facts. Even if engineers attempt to deal with design problems above or below their real area of expertise, few will be able to deal with all the subtleties and ramifications of their design decisions. Thus, the expected advantages of letting clever designers handle more of an overall system problem may well be offset by performance limitations induced by their lack of expertise in some areas.

Because there are so many different ways engineering workstations can be employed, and so many ways work can be divided, the manager contemplating a first purchase must decide early in the acquisition process exactly who the intended users are to be, and what is the expected result of their transition to new design technology. Failure to consider the user's limitations and preferences will almost always result in buying equipment not well suited for the job, with unfortunate financial and human consequences.

Finally, when considering the purchase of engineering workstations, it is crucial to realize that workstations may change the way organizations and people work. Engineers who never would have considered detailing part drawings, hand-drafting schematics, or laying out integrated circuits may find that they are very comfortable with these tasks on a workstation; many engineers exhibit considerable pride in their ability to deal with all aspects of a design, feeling that they do the job better than the traditional support people could do. This effect may reduce the organization's need for, or use of, technicians and drafters.

Conversely, some users will find that they do not like the online environment with its subtle pressures to take on all aspects of a design job (some of which are regarded as menial and degrading by many engineers); such users may resent the workstations and avoid their use entirely. Managers must be aware of these human problems and carefully allow for their impact when planning a workstation purchase. Systems must be sold to engineering departments with the same care and thoroughness that they are sold to upper management.

2.5 Beware the Vendors

It would be impossible to publish a complete or current list of all the vendors of engineering workstations in a book such as this one.

Appendix A lists many of the prominent suppliers as of late 1984, but new vendors appear, and others disappear, almost daily. Readers who plan to purchase workstations are advised to consult one of the many periodicals and newsletters in the field for an up-to-the-minute listing of vendors and products [2-4]. The fact that vendors come and go with such rapidity reflects the newness and the excitement of the engineering workstation business. However, it also suggests to the cautious buyer that there are substantial risks involved in choosing a vendor.

Fully-integrated engineering workstations of the type becoming common in the electronics industry, perhaps more than any other CAD/CAM product, represent a commitment and a partnership between vendor and user. Since, ideally, all the tools an engineer uses are integrated into the workstation, it is frequently not practical to use tools which are not supplied by the workstation vendor. When you evaluate the products of several vendors, you are likely to find that, for instance, engineers like the schematic drawing software offered by Valid Logic Systems Inc., but prefer the analytical capabilities offered by Mentor Graphics Corporation. Since each vendor's tools are unique to that vendor's hardware and database, it is usually impossible to mix and match. The user almost always has to choose the vendor which has *most* of the tools needed and the best *overall* capability, rather than buying the best of everything from multiple vendors.

Given that most buyers will probably have to purchase all their tools from a single vendor, it is obvious that the vendor must be reliable and stable. It is worse than embarrassing for your workstation vendor to go out of business; it may cause *you* to go out of business. Unsupported software and hardware quickly become useless. No company should depend for its livelihood upon the products of a departed vendor. As one frustrated engineer observed at a recent panel discussion, "Where do you go to get your Edsel fixed?"

Unfortunately, most workstations are built and sold not by established computer companies or CAD/CAM companies, but by "start-ups"—entrepreneurial companies founded to market workstations. Several of these companies have gone from their founding to revenues of $20 million or more in the space of a year, and on the success of one product. None of them are more than three years old at this writing.

Buyers of engineering workstations must be extraordinarily careful in their vendor selection. It is of the utmost importance that first-time buyers investigate not only the product, but the company they are buying. Insist on a home-office visit; meet the principals of the company; examine their financial position and business plan; talk to several of their existing customers and, if possible, their financial backers; take nothing for granted. It is much easier to demonstrate technical wizardry than it is to support a customer base.

As this book is being written, large established companies are beginning to enter the workstation market. However, it is too early to determine what impact they may have. Historically, small companies have provided the fundamental innovations in new markets such as this one, and large, older companies have been only marginally effective in competing (with the exception, of course, of IBM's entry in the personal computer market; IBM is an exception to almost all generalizations!). Potential buyers should thus be aware that the security of dealing with vendors like Hewlett-Packard, Tektronix, Control Data, and IBM may well be available by the time this work is published. Nevertheless, newer companies like Daisy Systems, Autodesk Inc., Skok Systems, Valid Logic Systems, and Mentor Graphics will probably continue to offer the most innovative products for some time to come. Potential buyers should not eliminate any vendor, no matter how small, based on this discussion alone. In the chapter on acquisition of engineering workstations, the subject of selection and qualification of vendors will be covered in much more detail

References

1. M. T. Fuccio, CAE Systems Based on the IBM Personal Computer, *VLSI Design Magazine*, June, 1984, p. 60.
2. CAE Systems: A Status Report, *VLSI Design Magazine*, June, 1984, p. 44-49.
3. CAD/CAM Alert, a newsletter published by Management Roundtable Inc., 824 Boyleston St., Chestnut Hill, MA 02167.
4. Workstation Alert, a newsletter published by Management Roundtable Inc., 824 Boyleston St., Chestnut Hill, MA 02167.

3
HARDWARE

The computer hardware for engineering workstations comes in two broad classes: generic and specialized. Each approach has its proponents and detractors, its features and its liabilities. This chapter will present in general terms the approaches taken by a number of important vendors. The use of specific vendor equipment and features for illustrations is a convenience for both the reader and the author. None of the following material should be interpreted as an endorsement of a particular vendor or approach.

The so-called industry standard, or generic, hardware approach is typified by the use of a general purpose computer, such as the Apollo or Sun Microsystems computer families, by such vendors as Mentor Graphics Corporation and Auto-Trol Corporation. The second, the so-called application-specific hardware approach, is typified by Daisy Systems, Cadnetix, and Valid Logic which build their own complete computers with specialized modules and subsystems.

There are many good technical and marketing reasons for these two hardware approaches. To the customer, the real issue is the fear of becoming committed to a particular vendor by a closed system, one on which only the vendor's software can execute, or which locks in the customer because of a complex or inaccessible database.

Becoming committed to a vendor can be a mixed blessing. The company contemplating such a commitment must weigh the consequences of a major change in the vendor's ability to support the product in later years. Many of the companies producing workstations are very new, and many are sure to disappear as a result of the natural selection of the marketplace. Those which survive and become highly successful frequently find that they have not developed adequate staff and resources to support a huge, demanding customer base. Either way, the customer who is tied to a single vendor for workstations may have to provide much future support for his own systems.

Less dramatic than bankruptcy or overwhelming success, but potentially as damaging to the customer, is the possibility that the vendor has not made plans for the long term in his architecture. A vendor which has committed to special hardware must plan for the evolution and refinement of that hardware, as well as the software. All computer systems evolve over their life cycles; will a system based on proprietary hardware evolve as rapidly as a system based on generic hardware? Or will it grow toward a niche market that is not relevant to a particular customer's needs: *your* needs?

Vendors of application-specific hardware make a compelling case for their decision to use nonstandard computers and specialized processors. These companies view the design of a workstation as a total systems problem, and try to partition the system into modules dedicated to subtasks (a process which is also central to the *use* of workstations). Either hardware or software (or both) may be applied in the solution of a particular systems problem, but hardware is usually the best answer when high performance is vital. Thus, for tasks such as schematic editing and word processing, for project management and database management, general purpose hardware is usually the most cost-efficient answer. However, for simulation and shaded image generation, specialized hardware can often provide orders of magnitude better performance than software running on general purpose hardware [1].

Vendors of systems based on general purpose hardware can also make a case for their viewpoint [2]. One good argument is that customers can often buy "portable" software applications, and use computing resources their companies already have in place; it may well be possible to buy software that is compatible with existing hardware. Customers are told to partition their tasks by computing power. If you already have a mainframe, try to find a workstation architecture that lets you use this resource easily. If you have a superminicomputer, particularly one of the popular VAX™ family from Digital Equipment Corporation, many vendors offer add-on software which can be run on it. This approach may offer the most cost-effective way to get started with workstations.

Whatever approach you choose though, it is wise to think through the implications of these two fundamental classes of systems and the tradeoffs they imply for your company. Whether the hardware is off-the-shelf or specialized, the user must determine whether his systems should stand alone, share a large processor, or operate as a distributed network. These decisions are not as easy as they may appear at first glance. Many capable engineering workstation systems are based on personal computers identical to those found at the local computer store. Some of these systems feature simulation, powerful graphics editors, and remote communications capabilities. In large companies dealing with multiple-engineer projects, the facilities of a distributed network may be attractive. In still other companies, the need for very large simulation and postprocessing operations may necessitate use of a mainframe; for these companies, a few additional graphics terminals and some well-chosen software may provide the optimal engineering workstation.

3.1 Workstations Based on Personal Computers

In this section, the term PC is used to mean personal computer, and not necessarily the IBM product of that name. However, the engineering community seems to have settled on the IBM Personal

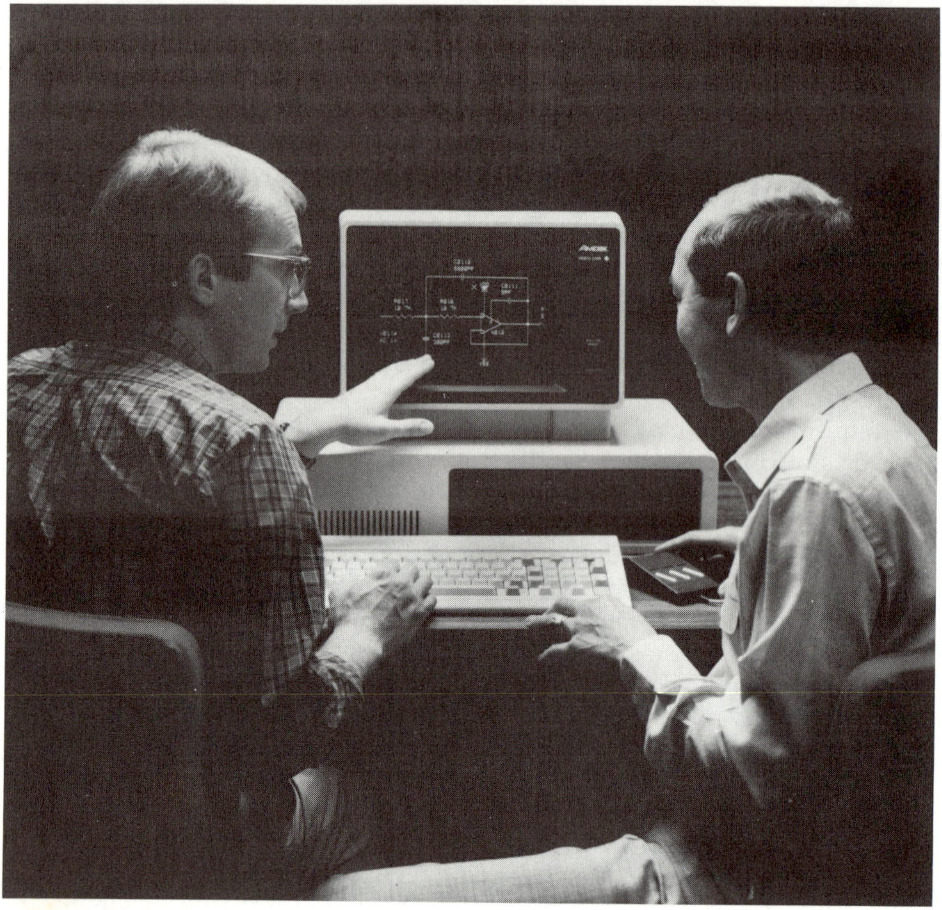

Figure 3.1 The PCAD™ system is a schematic capture workstation built by Case Technology Inc. and marketed by Control Data Corporation. It is based on the IBM Personal Computer. (Courtesy Control Data Corporation)

Computer (and compatible systems) as its standard vehicle for low-end engineering workstations, so the terms may indeed be synonymous at times.

Most of the current PC-based electronic systems support schematic capture graphics similar to the larger systems (Figure 3.1).

Systems for mechanical and architectural design almost universally feature drafting editors similar to those found on turnkey CAD/CAM systems. Two dimensional packages are the most common, but 3-D and even solid modeling packages are available on PCs.

Despite the wide availability of PC-based workstations, there are typically significant limitations in terms of database size graphics flexibility, and resolution. Packages sold as mechanical design systems frequently are simply mechanical drafting systems, capable only of producing drawings on a locally attached plotter. Some products recently introduced also advertise limited finite element modeling capability; these systems are of very limited use unless a larger computer is accessible for performing analysis.

Similarly, some of the electronic systems support logic simulation, and some permit limited analog simulation. However, all support interfaces with external simulators and layout systems in recognition of the need to perform these tasks on a larger computer. Thus, the first lesson to be learned by users of PC-based engineering workstations is that these systems rarely are capable of doing the whole job.

Many of the PC-based workstation products require that special hardware be added to the basic computer. Some simply require a higher resolution monitor than the PC manufacturer provides, while others require a special graphics driver card. For instance, Chancellor Computer requires a high resolution monitor, providing 1024 by 768 pixel resolution (a pixel, or picture element, is the smallest resolvable point on a screen). Futurenet Corporation, on the other hand, requires a special graphics controller which increases the resolution of a standard monitor to 640 by 360 pixels; the controller is sold with the Futurenet softnet software as a package. Autodesk Inc. requires graphics display upgrades to the many types of PC their AUTOCAD package runs on, but can use many different types of controller.

While most PC-based workstations are offered by small companies as standalone units, an interesting recent trend is the marketing of these small stations by larger companies as low-cost, compatible terminals to their full-function systems. Control Data Corporation, for instance, offers PCAD, a fairly typical PC-based electronic schematic capture system. The unique feature of this system is that it provides all the necessary tools to interfere with simulators and other programs on CDC'S mainframe network, including netlisters, syntax checkers, and communications programs. Using PCAD, the engineer can perform functions well suited to a small machine with confidence that his runs on the mainframe will not fail because of interface and setup problems. MacNeal Schwendler, a software company well known for its MSC/NASTRAN finite element analysis system on large computers, now offers a compatible product called MSC/PAL designed to run on a PC. Engineers can develop models for small parts and assemblies on the PC at low cost and then employ the more powerful NASTRAN system for stress analysis and heat flow on a mainframe. Designs developed on the smaller system are transportable to the mainframe-based software through conversion software provided by the vendor.

Racal Redac offers the REDLOG™ system based on the IBM Personal Computer XT with a special graphics board. REDLOG features elaborate graphics capabilities, with pop-up menus, rotating and "dragging" symbols, and "rubber banding." The software environment is compatible with the company's larger systems, supporting the same hierarchical design style and the ability to coordinate multiple-person projects.

Daisy Systems Corporation offers the Personal Logician (Figure 3.2), a low-price version of its Logician™ workstations, which utilizes an IBM Personal Computer rather than the more expensive (and powerful) Logician hardware. Like PCAD, MSC/PAL, and REDLOG, the Personal Logician provides tools unique to a particular vendor's software environment, providing the user a range of capabilities from low-end to high performance within a single product family.

For those considering purchasing workstations based on personal computers, a major concern is performance. A good rule of thumb for electronic designers might be: "Only simulate a circuit that you can see on one normal screen" [3]. Most manufacturers and users concur that analog simulation on even the best PC is limited to about 100 transistors, and that logic simulation without the aid of special processors is limited to around 100 gates.

However, remarkable performance is sometimes extracted from modest systems through the addition of relatively inexpensive hardware. For instance, Futurenet Corporation's DASH-CADAT system, an electronic design and simulation system, has been demonstrated to run large fault simulations as fast as far more

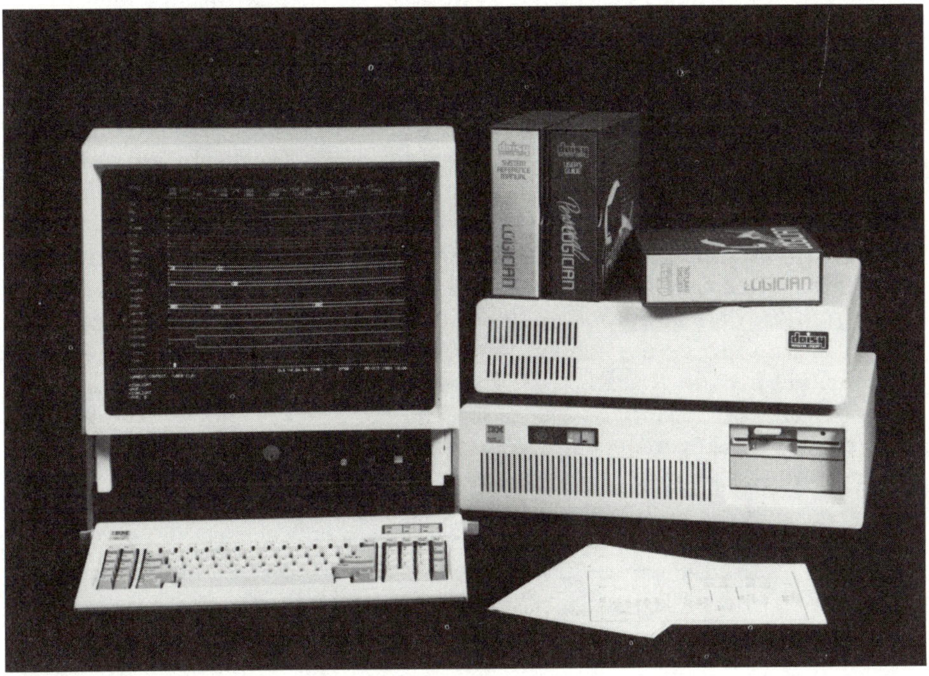

Figure 3.2 Daisy Systems Corporation has developed the Personal Logician™, a low cost version of its popular Logician™ system that uses an IBM Personal Computer rather than the vendor's proprietary computer. (Courtesy Daisy Systems Corporation)

powerful workstations through the addition of a special processor on one board of an ordinary IBM Personal Computer [4].

3.2 Workstations Based on Mainframe Computers

There are many vendors supplying engineering workstation software which runs not only on personalized systems but on large scale computers. In many cases, the software and user interface process, and perhaps even the file system, are identical to those used on smaller machines. The advantages cited by vendors of these packages range from software portability to computational power to economy.

The Control Data PCAD system, discussed earlier in the section on PC-based systems, consists of an IBM or Columbia PC and software for schematic editing, plus a set of "ISD Utilities." The system's unique feature is that it is a cooperative effort between a mainframe vendor/timeshare house (Control Data Corporation) and an engineering workstation software vendor (Case Technology Inc.). Although the workstation environment is provided, it is really a front end to a mainframe, with utilities and interfaces provided for a specific mainframe environment and the tools Control Data offers. Thus this system in a sense represents a mainframe based system, not a standalone engineering workstation. The ISD Utilities include a tailored asynchronous communications package, syntax checkers for the ASPEC, LOGIS, and SYSCAP simulators, and netlisters for these packages plus TEGAS, SPICE, and AUTOMATE-80. The mainframe system vendor thus provides a complete environment of design tools for the user on a small, economical workstation; substantial revenue from usage of Control Data's mainframe timesharing service is an inevitable result.

The Cadtec series 8000 electronic design system uses the VAX™ line of super-minicomputers with Orcatech graphics terminals. Cadtec also offers lower cost documentation/office stations for project management. The use of a shared computer allows a single, central database for large projects; configuration control software for project managers is a major feature of the system. While workstations can be configured by the user to operate on private copies of data, upon release to the central system designs must fit the data structure defined for the project. Systems of this sort are easier to implement on a centralized, mainframe hardware architecture than on totally independent workstations [5].

The CAE-2000 software package, a product of CAE Systems Inc., is available either on mainframes (the IBM 4300 series or the Digital Equipment VAX series) or on networked small systems (the Apollo, Sun-2, or MicroVAX). The unique feature of the CAE-2000 system is that the software environment, the screen presentations, even the database, is transparently portable across all the supported machines.

The Computervision 3000 and 4000 series workstations may be integrated with a series 5000 mainframe system. While the 5000 is not strictly a workstation product, its primary function is to manage large data bases created and edited on the smaller workstations in the product line. The products represent a hybrid hardware approach, and a growth strategy for large customers, an approach currently unique in the industry.

3.3 Distributed Systems and Architectures

Most engineering workstations employ some sort of interconnection method that allows engineers at different stations to communicate and share data. However, there are a great number of ways in which this interconnection may be accomplished, and a great variety of performance and operational differences between them.

Briefly, there are three basic types of connection which may be used. The most common among traditional CAD/CAM systems is the shared host approach, the least flexible means of communication used for engineering workstations. Most of the PC-based systems discussed above can be connected to a shared host for simple communications; these connections usually consist simply of a low-speed serial line, such as is commonly used to attach terminals to a computer. In these systems, there is generally no convenient way for users to share resources or design data other than by file transfers through the central computer.

The bus and ring schemes employed in many workstation systems are much more capable and flexible than serial links. A bus system employs a single cable with "taps" at each station. There is no master, and each station must compete with all others for access. A ring scheme is synchronous, and any station may transmit when its turn comes up. These systems will be discussed below by describing typical implementations in real systems.

3.3.1 Bus Network Systems

Many vendors of Engineering Workstations utilize the Ethernet™ system (Xerox Corporation) to interconnect workstations. Among these are special-purpose hardware suppliers like Daisy Systems (Figure 3.3), Valid Logic, and Metheus-CV Inc. (Figure 3.4) as well as companies which add software to general purpose computers from Sun Microsystems, Masscomp, Digital Equipment Corporation, and many others.

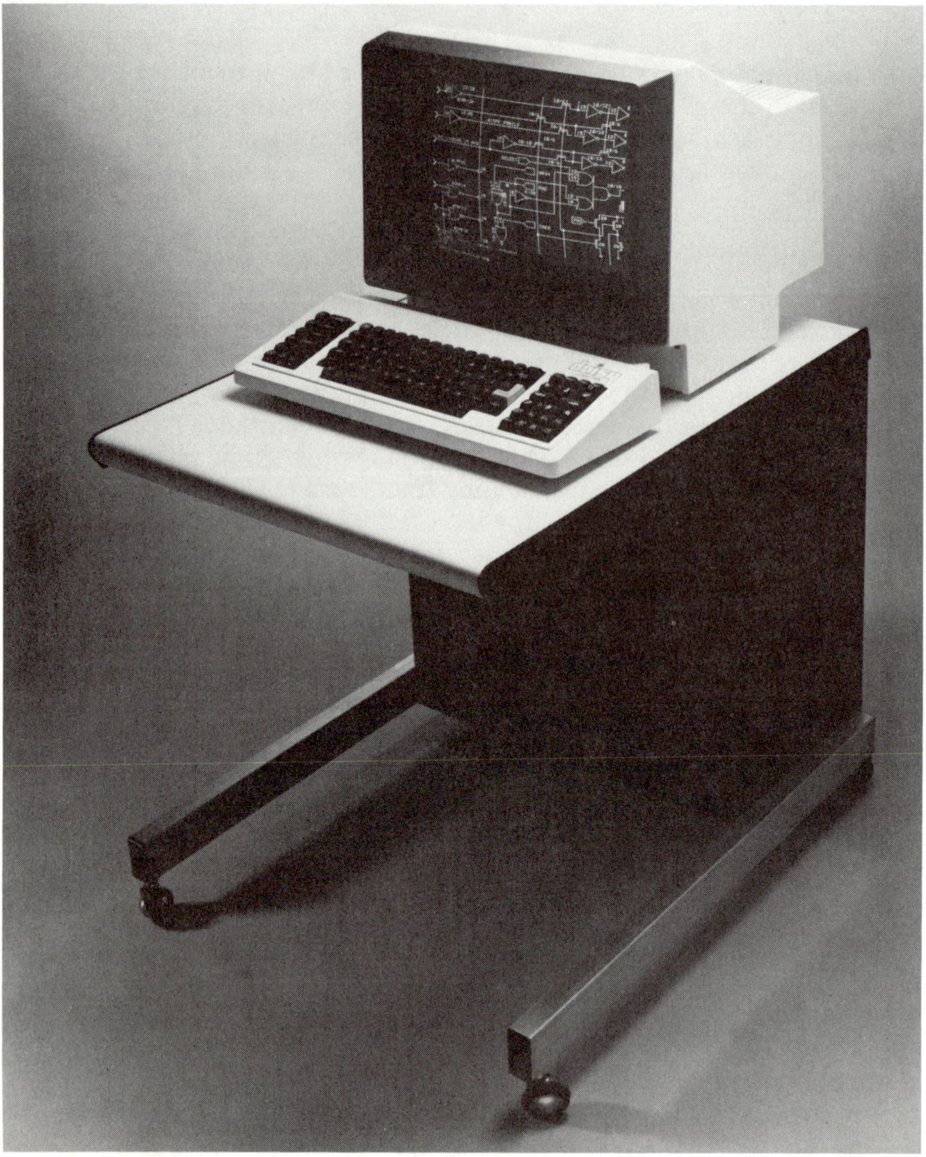

Figure 3.3 The Daisy Logician™, one of the first engineering workstations available, can be interconnected to other Logicians and systems using Ethernet™. (Courtesy Daisy Systems Corporation)

Figure 3.4 Although the Metheus-CV Lambda 750 workstation is a highly customized system, it communicates with other MCV systems through Ethernet™. Use of this common interface allows the Lambda 750 to communicate with other vendors' systems as well. (Courtesy Metheus-CV Inc.)

Ethernet systems are characterized by a single cable connecting each station, and the lack of a master node. Ethernet is a "Carrier-sense, multiple access with collision detection" system (CSMA/CD); this formidable title describes the important features of the way such a network operates. Briefly, the network operates like this: A station which has a message to send builds a packet of information which contains the address of the intended receiver, the data to be sent, and the appropriate error-correction codes. The sending node listens to the bus (carrier sense), and waits for a time when there is no data being sent by another node (multiple access). When it detects a quiet period, the sender broadcasts its packet of data onto the network.

As it sends its data, the sender continues to listen; in the unlikely event that another node begins broadcasting simultaneously, the node in question will sense the collision between its data and the other node's by comparing the data it sent to the data it hears on the network. If the two (or more) senders hear the collision, both will enter a state called "backoff." In this state, each sender waits for a randomly determined time interval and then re-initiates the send sequence.

Assuming the network is not heavily loaded, the random occurrence of a collision is quickly corrected by the random backoff and retransmission. Problems only occur on such a network when the amount of traffic increases to about 50 percent of the theoretical capacity of the cable. Beyond this level, there may be a marked degradation in the throughput of the network.

The above (admittedly oversimplified) description of the operation of an Ethernet only describes the physical transmission of packets of information. Indeed, the IEEE standard only defines the protocol of an Ethernet to this level; manufacturers of systems which use Ethernet must define for themselves many details such as the arrangement and formatting of data in packets, the types of data which can be communicated, the means by which the network is used to implement file transfers, and many other details. Many developers support higher level *de facto* standards such as Telecommunications protocol/Internet Protocol (TCP/IP, the scheme developed by the U.S. Department of Defense for interconnecting the systems of many vendors) and XNS™, the simpler protocol defined by Xerox Corporation for communication between office systems. Systems which support these standards can usually at least communicate with each other, although such communications may be limited to simple messages and perhaps file transfers.

Bus networks like Ethernet have several advantages over ring networks, and several disadvantages as well. The Ethernet cable need not form a continuous loop, permitting somewhat more flexible routing of cable and location of equipment than a ring network. Use of an Ethernet network frequently makes possible the integration of hardware and software from a variety of vendors, thus avoiding overdependence on a single vendor.

The major disadvantages of Ethernet lie in its complicated restrictions on total bus length, bus segment length, number of nodes, relationship of branches, repeaters and links, etc. (Details of these considerations are beyond the scope of this book, but may be found in any good text on data communications, a few of which are listed at the end of this chapter.) An additional problem is the unpredictability of throughput as network load varies, a fact that has prevented some workstation manufacturers from implementing transparently distributed systems comparable to those which use ring networks.

3.3.2 Ring Network Systems

Ring networks, like bus networks, use a single cable between nodes and have no master node. However, they differ significantly from bus networks in many ways. The ring network, as its name implies, consists of a cable that circulates through all nodes and returns to the beginning; an Ethernet or other bus system is open, with terminators at each end of its cable to absorb each message.

There are several types of ring networks in use today. The details of their implementation and their performance are not important to the understanding of their principles (and thus, are omitted here to keep with this book's goal of minimizing irrelevant detail). Readers with an interest in the subject are referred to the many good books on the subject [6,7].

As is the case with bus networks, ring network communication protocols are defined in many layers, all the way from the physical characteristics of cables and connectors to the rules for communication between applications programs. However, since there is no accepted standard for ring networks, even at the lowest levels, the trend is for each manufacturer to implement its own unique network strategy to exploit its hardware and software to the fullest extent possible.

The most popular ring network for use in engineering workstations is the DOMAIN™ network used by Apollo Computer Inc. In this network (technically known as a Newhall Loop [6,8]), a "token"—a single small packet of information to which messages are attached by the computers in the ring—is passed from node to node around the loop. Only one token, and thus only one packet of information, can be circulating at a time. When a node wishes to send a message, it waits for the token to arrive at its network input; as the token passes, the message is attached to it and sent on to the next node. Each node in turn examines the message and passes it on, until it reaches its intended destination. The receiving node passes on the token, but absorbs the message into its buffer, forwarding an empty token available for use by another node.

Networks of this type, like the other architectures discussed, have strong and weak points. Their major strengths are that a single node cannot dominate the network, and that messages may have essentially any length. The major theoretical weakness of the ring architecture is that a node failure can interrupt the entire network. The DOMAIN system, and most other ring network systems, has a failsafe mechanism in the ring interface circuitry, which automatically inserts a passive link in place of a failed repeater.

A major advantage of a ring network over Ethernet is the consistency of message passing time. In a system such as Apollo's DOMAIN, the time is consistent and short, allowing very extensive resource sharing and virtual memory operations across a busy network. These capabilities result in highly-integrated distributed system which feature resource sharing and effortless use of communications. In a ring network, the user literally may not know or care where his files and programs are located, since the network takes care of all the details of communications in a fashion invisible to the user. A busy Ethernet network, being asynchronous and subject to delays of random length, may have very inconsistent throughput; this characteristic makes it difficult to implement highly-interactive systems that depend on the network.

Thus, the particular strength of the Apollo system (Figure 3.5) is not in the network architecture so much as in the software environment provided. The native capabilities of the system for file transfer, peripheral sharing, and interprocess communication are used to advantage by many workstation vendors who base their design systems and libraries on the Apollo distributed architecture.

3.4 Specialized Hardware

Several prominent vendors have adopted a strategy of building specific hardware for their workstations in addition to writing software. The issue is quite controversial, with articulate spokesmen on each side presenting convincing arguments in the print

Figure 3.5 The Apollo DN-550 is a moderately priced color graphics system used by several workstation vendors. It, like all other Apollo systems, communicates through the proprietary DOMAIN™ token ring network. (Courtesy Apollo Computer Inc.)

media and at industry forums regularly [1,9,10]. The following material is included because, to the first-time buyer of engineering workstations, the strengths and weaknesses, the advantages and disadvantages of specialized hardware may not be obvious.

Specialized hardware can be used to attack unbounded problems, those which are likely to grow rapidly in the future. To illustrate, consider a task like word processing which is bounded: the vocabulary and the potential size of documents are not changing significantly. Adding to the sophistication of software and hardware supporting this kind of task has had little marginal benefit on productivity. Tasks like finite element analysis and analog circuit simulation are essentially unbounded for the foreseeable future. Engineers are currently constrained in the amount of these resources they can bring to bear on a problem by the computing power they are able to obtain. Even hardware that increases the efficiency of these tasks by many orders of magnitude (a feat essentially impossible using software solutions) will not behave like the word processing example. The size of engineering problems will continue to increase without bound, and the amount of analysis required will follow. For this reason, many workstation vendors apply hardware solutions in the areas where they feel they can obtain the greatest productivity gains [1].

The integration of tools is a critical issue to users of workstations. The vendor of specialized hardware can design hardware solutions to be fully integrated and compatible with his software; the vendor who resorts to software solutions only is at the mercy of its hardware vendors. Proponents of application specific hardware feel it is inevitable that today's software-only vendors will be forced by competitive pressure to begin providing some hardware solutions. When this happens (as with Mentor Graphics' use of Zycad logic simulation accelerators), the integration of the new capabilities may compromise the vendor's ability to maintain a clean interface between tools.

The case for general purpose computers can be equally compelling. The vendor need not concentrate a substantial portion of its resources on hardware development and support; ideally, all resources are concentrated on tool integration and development, customer support, and refinement of software. Such a company theoretically can focus on the application and the customer more effectively than can a vendor who must also develop and support complex hardware.

A third party supplier of standardized hardware will typically offer a range of performance and feature levels, representing many points on the price-performance curve. The application-specific hardware vendor, on the other hand, will probably offer only a single central processor, a single simulation engine, and so on. The user of general purpose computer-based workstations can draw upon a rich variety of software available from sources other than the workstation vendor: simulators, interfaces, graphics programs, office automation tools, and an open-ended host of other software are largely unavailable to the user locked in to a specialized processor.

The essential paradox of the argument over specialized hardware is that the very tools the engineering workstation industry is producing are making it possible to design and build specific hardware for a task very economically. As recently as five years ago, building an all-new specialized processor for, say, logic simulation, was an enormous task. However, with a high performance engineering workstation network, a small team of engineers can produce a product like Valid Logic's REALFAST™ or Daisy System's PMX™ in very short order. Such tools, in fact, accelerate the pace at which they can be developed and succeeded by better tools.

3.4.1 Simulation Accelerators

Accelerators are special purpose computers which perform a particular computational process much faster than a software solution. For example, the ZYCAD LE-1000 series, the Daisy Systems Megalogician™, and the Valid Logic REALFAST™ are engines for logic simulation which run several orders of magnitude faster than software logic simulators.

Logic accelerators potentially solve the problem of throughput for large simulations; however, they depend for their success on hardware models that accurately represent the logic primitives used by designers. The ZYCAD machines, for instance, represent all logic as networks of three-input gates which perform Boolean logic functions. Since no one actually implements logic designs this way, a compiler must translate the designer's logic into the appropriate models for simulation. Many designers of integrated circuits feel that this translation introduces a level of abstraction that makes accurate simulation impossible. Other designers, particularly those who use catalog logic functions in their designs, feel that compilation is quite acceptable and introduces no significant inaccuracies.

The Daisy Megalogician represents a different approach than the ZYCAD LE-1000. It mimics exactly the primitives and logic states which Daisy's software simulations use, thus avoiding the compilation and abstraction steps of the ZYCAD system. Megalogician supports logic, fault, and timing verification. Similarly, the Valid Logic REALFAST system utilizes high-level primitives, making its use transparent to the user. The only perceived difference between using REALFAST simulation or the vendor's SCALD™ Logic Simulator is throughput.

3.4.2 Graphics Accelerators

The generation of complex images on a workstation screen is a task requiring an enormous amount of computation. Each data item in the model being displayed must be scaled, transformed to the proper viewing angle, clipped to fit the screen, checked against many other elements to eliminate hidden lines and surfaces, and transformed from a line representation to the appropriate pixels. Each pixel must then be set to the correct intensity and color value depending on the lighting angle being simulated. These operations, although complex, are highly structured and repetitive, and thus lend themselves well to implementation in hardware. The result of special hardware for image generation can be very dramatic. Early solid modeling systems frequently required hours of computation for generation of a static image on the screen; some of today's compact workstations can generate similar images in seconds, and can transform them to different viewing angles and lighting perspectives nearly instantaneously. Such capabilities make possible workstations for architectural renderings and for mechanical design of complex mechanisms.

One popular graphics accelerator, the Weitek Tiling Engine™, is sold as a set of chips for use by system builders. This system is used by several manufacturers of small computers to enhance graphics performance in workstation products. Other manufacturers design their own proprietary chip sets to produce extremely high-performance graphics systems like the Silicon Graphics Inc. IRIS system.

3.4.3 Algorithms in Silicon

A particularly exciting hardware development, not yet on the market, is the attempt by Manufacturing and Consulting Services Inc. (MCSI) to implement its entire drafting software package as a chip set. By the end of 1985, the company plans to have all its mathematical processes and algorithms for mesh generation, numerical control, and solid modeling "cast in silicon." This effort is being pursued jointly by MCSI, Weitek, and Hewlett-Packard, and could result in engineering workstations of extraordinary high performance by the latter part of the decade.

3.4.4 *Physical Modeling Systems*

Physical modeling systems are an innovative class of specialized hardware aimed specifically at designers of electronic systems. These devices began to appear in early 1984 and represent a major new capability for some engineering workstations. Valid Logic's REALCHIP™ was the first system on the market, followed by Daisy's PMX (Physical Modeling Extension). Mentor Graphics has very recently introduced their Physical Modeling Library, and CAE Systems Inc. has introduced Turbochip™. Several other vendors are known to be working on systems of this type, and some of them may be on the market by the time this book reaches print.

The REALCHIP system consists of a set of reference elements on adaptor boards, mounted on device boards that plug into a mother board. Altogether, there may be up to 64 reference elements in a REALCHIP system. Each reference element is a physical component such as a microprocessor; it has a corresponding library entry in the SCALDsystem™, consisting of an outline drawing and a device definition text file. Given these three components (the physical part, the pictorial representation, and the definition file), a user may treat the REALCHIP model exactly as if it were a conventional logic primitive, using it in multiple instances throughout a design if desired. The operation and behavior of the workstation's tools are identical, whether the user deals with physical components, with software models, or a combination: the only difference is speed. Combined with the manufacturer's REALFAST simulation engine, the system can simulate systems as large as several million equivalent gates in reasonable times.

There are many advantages to the use of a physical modeling system. The most obvious is that the user need not develop complex software models of standard components like microprocessors in order to design systems around such components. Less obvious is the ability to simulate a custom component like a gate array prior to building the component, and then to substitute the real thing into a system design later and validate its performance. Finally, vendors claim that the speed of the physical component is independent of its simulated speed; in other words, a low-performance preproduction version of a chip may be used with a device definition that declares it to be a full-speed component. The simulator thinks it is dealing with a full-speed device, while the REALCHIP hardware operates the actual component at its degraded speed. This capability allows designers to have a headstart in designing new systems, before full performance chips are available.

The physical modeling system technology is still very young, and may not realize its full promise for several years. Among the many problems inherent in this type of system is the speed difference between the software models on the workstation and the hardware seen by the system. Potential users of these systems should be prepared to ask many questions and to demand demonstrations of stated capabilities. For instance, if dynamic parts are being used, how are states maintained on the physical hardware when the user pauses the simulation for a look at intermediate states? Is it necessary to employ a logic accelerator in order to use the physical modeling system? Are the delays and clock rates used in simulation typical, worst case, or best case? Does the temperature of the physical chips have any effect on the simulation? All these, and many other questions, are currently being asked by the vendors and their pioneering customers. Readers should be aware of the controversy surrounding physical modeling systems and investigate their ramifications thoroughly before accepting a vendor's opinion of their usefulness.

3.4.5 Logic Analyzers

Logic analyzers, a highly-specialized type of digital oscilloscope, have been available as benchtop debugging tools for years. Recently, specialized logic analyzers designed specifically for use with engineering workstations have become available. One such system is the Mentor Graphics Hardware Verification System™ (HVS), a full-function logic analyzer which is integrated with the vendor's IDEA™ engineering workstation. The HVS may be driven by data patterns developed on the workstation from logic or fault simulation, and the results displayed or stored by the system. Since the user has access to all the other tools available on the workstation, the HVS provides many capabilities unavailable on conventional logic analyzers. For instance, the workstation provides very elaborate software tools for analyzing bus states in microprocessor systems. The system can display states in terms of processor instruction mnemonics, waveforms, bit streams, or even in special ways defined by the user. Patterns for driving the HVS may be derived from simulations run on the workstation, or from programs written by the user. Interaction with the HVS is through the standard screen interface used for other Mentor tools; schematics may be utilized to point at nodes the user wishes to examine in the hardware, in exactly the same fashion he interacts with a simulation.

Another system of this type is the HHB-Softron Chip-checker™. This system performs most of the same functions as the HVS, but is offered as a peripheral device that can be connected to a variety of different brands of computers and workstations. This product represents one of the many tools now being offered that is not intended for use with only one type of workstation, a very significant trend for users who wish to avoid vendor dependence in their system decisions.

3.5 Final Notes

In this chapter, we have only touched the surface of the complex world of engineering workstation hardware. As in so many areas, the systems targeted at electronic design applications are featured, simply because they represent the most technologically advanced systems, and the most widely available systems, on the market. Other disciplines will be influenced profoundly by similar developments in the near future. Graphics accelerators are vital to three-dimensional image generation for mechanical applications; finite element engines are sure to appear in the near future, offering mainframe performance at workstation prices. The field is changing so rapidly that it is impossible to cover or to anticipate all the possibilities. As stated several times above, the existence of engineering workstations is changing the entire face of the computer hardware industry, and drives the evolution of the workstations themselves. Readers should be aware of the rapidly changing nature of the hardware business, and acquaint themselves with the latest in technology prior to purchasing any vendor's product.

References

1. Lucio Lanza, The Case for Application-Specific Computers, *Engineering Manager Magazine*, September, 1984, p. 65.
2. CAE Workstations: A Perspective. Panel held at the IEEE Conference on Computer-Aided Design and Continuing Education in Engineering, September 12, 1983, Santa Clara, CA.
3. Mark T. Fuccio, CAE Systems Based on the IBM Personal Computer, *VLSI Design*, June, 1984, p. 53.
4. Special Report on Personal Computers in Engineering, *Electronic Engineering Times*, May 21, 1984.
5. A. S. Rappaport, Computer Aided Engineering Tools, *EDN Magazine*, September 15, 1983, p. 106.
6. Carl Tropper, *Local Computer Network Technologies*. Academic Press, New York, NY, 1981.
7. David C. Flint, *The Data Ring Main: An Introduction to Local Area Networks*. Wiley, Hayden, & Sons, Chichester, UK, 1983.
8. W. D. Farmer and E. E. Newhall, An Experimental Distributed Switching System to Handle Bursty Computer Traffic, *Proc. ACM Symp. Problems Optimization Data Comm. Systems*, Pine Mountain, Georgia, October 1969, pp. 1-34.
9. Ed Morgan, Is the Time Right for CAE Workstations? *Engineering Manager Magazine*, September, 1984, p. 58.
10. Steve Swerling, The Case for General Purpose Computers, *Engineering Manager Magazine*, September, 1984, p. 65.

4

SOFTWARE FOR ELECTRONIC ENGINEERING WORKSTATIONS

Electronic engineering workstations first appeared as products in 1981, although many of their capabilities had existed on mainframe-based systems and turnkey CAD/CAM systems for years. The thing which distinguished these systems most dramatically from their predecessors was their attempt to integrate all functions through elaborate human interface software. While engineering workstations, like other CAD/CAM systems, are based on discrete software packages for editing, simulation, formatting, and so on, many also have a "glue" which attempts to bind all the diverse engineering tools together into a cohesive whole. While prior systems required learning a different set of commands for each application, and perhaps learning to communicate with several different computers, the engineering workstation environment attempted to eliminate the need for specialized knowledge about the system itself. As a result, the systems are easy for a busy engineer to personally use. Direct use by engineers, as stated earlier, is the most important feature of these systems, and results in dramatic improvements in creativity and productivity.

Each manufacturer approaches the problems of design a little differently, and as a result, the software they supply differs. Nevertheless, there are certain functions nearly all suppliers provide in some form. The purpose of this chapter is not to describe in detail how all manufacturers systems work and how they differ; likewise, it is not intended to compare features and differences between the systems. Rather, the major applications found on today's engineering workstations will be described briefly, and supplemented by descriptions of a few particular packages that demonstrate the important concepts well.

4.1 Schematic Processing

The schematic diagram is the fundamental means of communication between electronic designer and workstation. This was the first function implemented on virtually all workstations, and indeed is the foundation of the engineering workstation business. Schematic editors attempt to answer two needs: data entry (to simulators and other CAD software) and documentation.

Data entry through a schematic has several important features for electronic or electrical designers. The schematic diagram is the most natural and precise means of describing a design, whether a communications circuit, a power substation, or a computer logic diagram. All design engineers in these fields understand the symbols and the meaning inherent in schematic diagrams, and all use them as their initial way of expressing a system of connections between components, whether or not they use an engineering workstation.

Most CAD tools that deal with circuits utilize some form of textual list as their input, usually a net list or a component list. Either of these forms of circuit description can be derived uniquely from a schematic diagram. Prior to the availability of engineering workstations, engineers desiring to use CAD tools had to encode manually the required lists from schematics they had sketched or drawn. This process is not only tedious, but is highly error-prone.

For instance, the engineer might omit a connection, or record a pin number incorrectly. In many cases, since simulations do not necessarily exercise all parts of a circuit, it may appear that a circuit functions perfectly. Later, perhaps only after a prototype has been built, mistakes made in the encoding process may be discovered. This example is not as far-fetched as it may seem: the first engineering workstations were designed to eliminate precisely this problem.

The second major function of a schematic editor is documentation. Production of finished drawings of a design, like encoding netlists, is tedious, expensive, and error-prone. There are many ways to err in this process, and all may remain hidden until much later, when faulty archival drawings or field service manuals may prove very expensive. For instance, when performing simulations, engineers routinely find mistakes in their preliminary designs, or implement minor changes; these changes are usually made in the netlist for the simulator, not in the original drawings. Despite good intentions and record-keeping, many such changes are never incorporated into the final drawings, or are recorded incorrectly.

Most engineering workstations automate the generation of net lists, thus removing the human factor and eliminating a major source of error. In addition, the generation of the netlist is usually somewhat hidden from the user, making it difficult or impossible to alter the list manually; rather, the user must make all changes *directly to the schematic diagram* using the editor, thus assuring that documentation and simulation are always in step.

It is not difficult, then, to see the benefit of schematic editors: the most natural representation of a design idea is used to provide error-free specifications for CAD analysis tools and to produce correct documentation.

4.1.1 Symbol Creation

A schematic diagram consists of symbols (representing components or functions) and connections (representing wires or other means of establishing electrical connections among components). Schematic systems usually provide two separate tools for dealing with these parts of the schematic.

A symbol editor provides the workstation user a means to create a library of parts to be used in his design. For each symbol to be used on a schematic, a graphical picture must be created. In

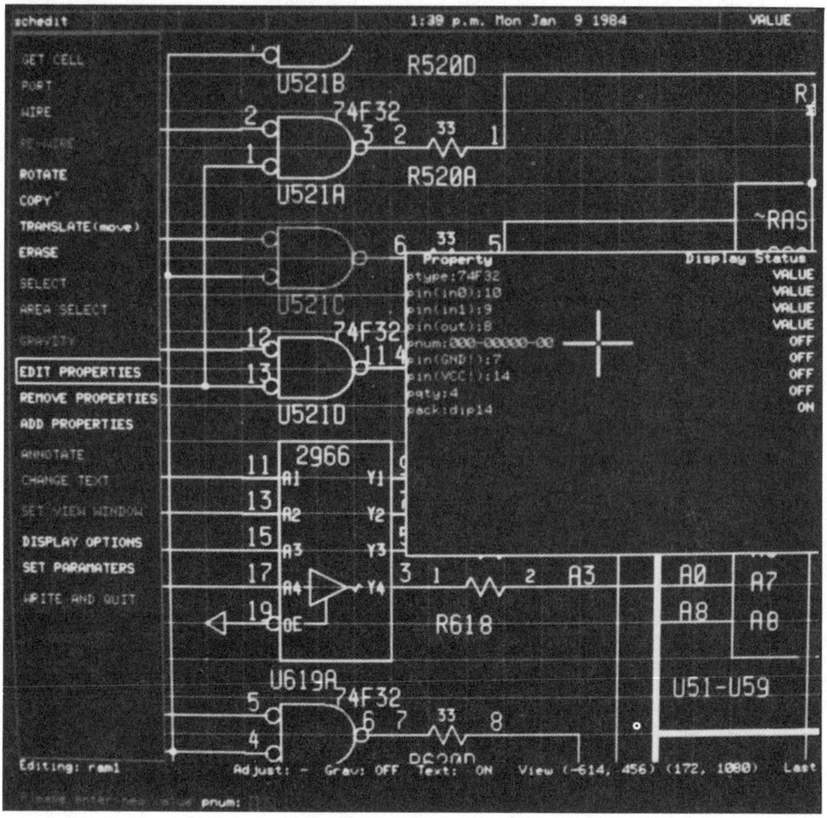

Figure 4.1 Symbols for use with a schematic editor may be supplied by the workstation vendor, or they may be created for particular applications by the user. Here, a user is adding special properties to a pin of a gate symbol. (Courtesy Metheus-CV Inc.)

addition, it is necessary to define the connection points to the device; this information is used by the CAD tools which must interpret the schematic to determine how the various symbols are connected to each other. Figure 4.1 illustrates the symbol editor screen display of a typical system (the Metheus-CV Lambda 750 workstation) during the addition of pin descriptions to a gate symbol.

In addition to the graphic symbol and the connection points, most symbol editors also allow the attachment of properties or attributes to the symbol. These nongraphic features (so-called because they are usually not visible on drawings or screen displays) supply information to other software used on the workstation. Typically, they provide information which, to a human observing the drawing, is understood, but not explicitly communicated by graphic symbols alone. For instance, a property attached to connection points may specify whether the point is an input or an output; another may specify the capacitance, or the drive capability of the terminal. Properties attached to the symbol as a whole may include part numbers for use in generating a bill of materials, or the name of an associated symbol (commonly used on logic gate symbols to refer to a separate symbol containing the gate's internal circuitry in terms of transistors and other fundamental components).

Although all engineering workstations for electrical and electronic design permit the user to create customized symbols, many users will never use this capability. Most workstation manufacturers provide libraries for common components and popular gate arrays. In addition, many manufacturers of components provide extensive symbol libraries for their products for the major types of workstations; vendors of gate arrays and other semicustom circuits often provide libraries of symbols and simulation parameters for their products as well. Readers intending to design their own symbol libraries should be aware that this is a very time-consuming activity. It is not unusual for an organization to spend six months or even a year developing customized libraries. During this time, it is unlikely that any design work can be done.

4.1.2 Network Editing

A network editor is a specialized graphics program which permits very limited manipulation of symbols, connections, and text. Although conceptually similar to the drawing editors common on CAD/CAM systems, those on engineering workstations usually have far fewer functions and far more constraints. For instance, many schematic network editors only permit lines to be drawn horizontally or vertically; text may be permitted only in the context of attributes like part numbers or signal names, and may be limited to only one or two fonts; company logos and title blocks may have to be created as special symbols, instead of free-form lines on a drawing.

The simple graphics capabilities of a network editor are deceiving. Its true capabilities lie in its knowledge of connections. If a line is terminated on (or near) a symbol connection point, the editor knows the designer meant to establish an electrical connection to the component. Frequently, the editor will snap the line to the nearest connection point and establish the connection; any subsequent movement of the wire or the component will not break the connection.

Some editors are clever enough to understand certain wiring rules, and will prevent users from connecting together more than one output in a net, or shorting together two nets which have unique names. Many editors will allow the symbol for a part to be moved, and will preserve all connections to the part, no matter where it is placed on the drawing. The drawing editor of a typical CAD system for drafting schematics, by contrast, will leave the wires exactly where they were drawn originally when a part is moved. The user must then move each wire and re-establish all connections manually.

4.2 Logic Simulation

Logic simulation is the most heavily used application on most engineering workstations. Designers of large integrated circuits, computer systems, control systems, and most other complex electronic devices today rely heavily on the ability of computers to verify the expected behavior of their circuitry. A logic simulator does not attempt to model in detail the voltages and currents in a circuit; rather, it deals only with a few states, particular voltage or current levels to which the designer assigns logical meanings. For instance, in many computer circuits, any voltage less than about 1.0 volt is considered to be a logical zero, and anything from about 2.1 to 5.0 volts is considered a one. Rather than deal with these detailed voltage levels, a logic simulator evaluates a simplified set of conditions and declares the state of a node to be a one or a zero. Many simulators also are able to identify ambiguous circuit conditions, like 1.5 volts in the example, and declare

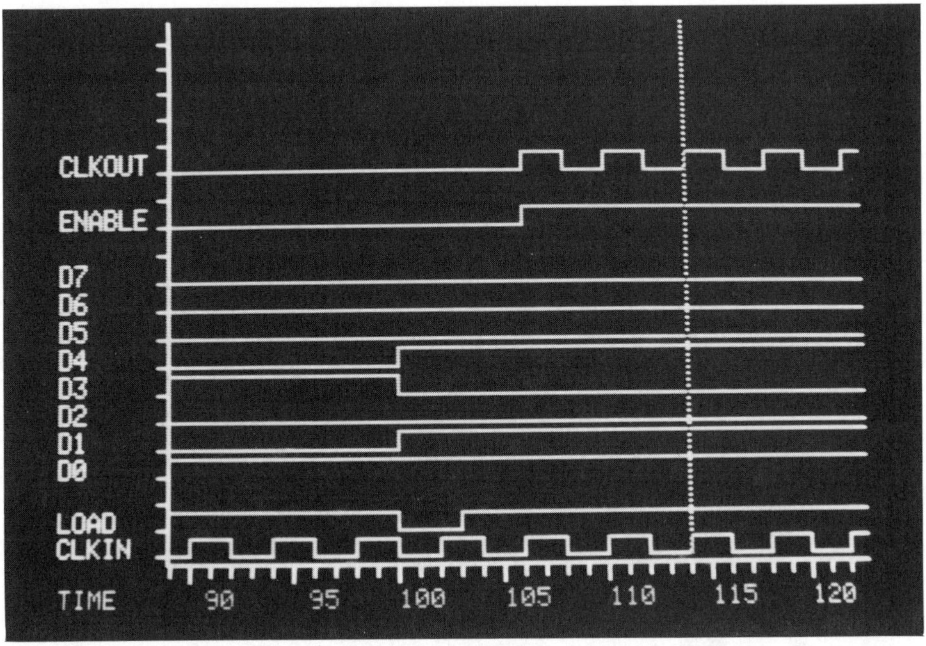

Figure 4.2 The LOGICPRO™ system from E/Z CAD Inc. provides a display of logic signals produced by the simulator in a form readily interpreted by the engineer.

these to be unknown or transitioning states. Using simplifications like those described, very elaborate digital systems can be modeled using computers with quite modest capabilities.

For instance, Figure 4.2 shows the screen display of the LOGICPRO™ package from E/Z CAD Inc., which runs on an IBM Personal Computer. The figure illustrates 12 different points in a circuit selected by the designer for monitoring. Notice that each wave exhibits only two states, representing simplified logical conditions like those described above. Such information can easily be displayed and understood graphically, and this type of presentation is common to nearly all electronic engineering workstations. Note also the vertical dotted line in the photograph; this is a cursor, posi-

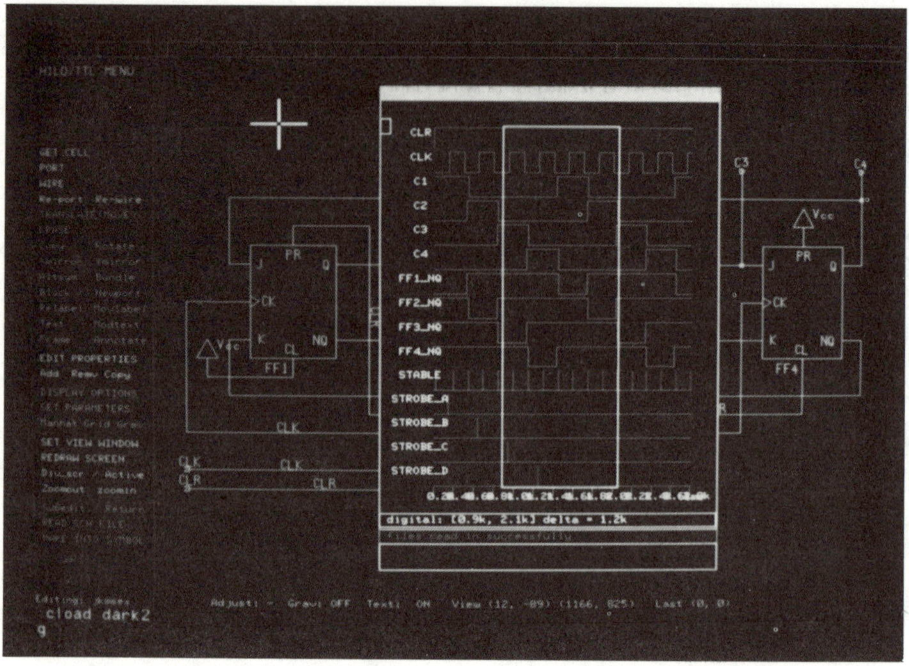

Figure 4.3 Multiple windows aid the workstation user in relating simulation results to the circuit being developed. (Courtesy Metheus-CV Inc.)

tioned by the designer, as an aid to examining the relationships of all signals at a particular time point of interest. Systems of the type illustrated typically require that the user select a group of signals to be monitored prior to running the actual simulation. Choosing additional signals, or changing other conditions of observation, usually requires that the entire simulation be run again.

Figure 4.3 illustrates a more elaborate simulator, running on the Metheus-CV Lambda 750 workstation. Notice in this illustration that the schematic diagram as well as the waveform display are visible simultaneously in different windows of the display. In this illustration, the user has run the HILO-2™ simulator (GenRad Inc.), and is interactively examining the behavior of various points in the circuit by pointing at them (with the workstation's "mouse" device). Systems of this type allow the designer great flexibility in evaluating the results of simulation runs. For instance, it is often possible to examine different groups of signals and different time periods without having to rerun the entire simulation.

4.3 Fault Simulation and Test Generation Tools

A fault simulator is a specialized version of a logic simulator and whose purpose is to determine whether a given set of test data will actually test all parts of a circuit. To the engineer unfamiliar with the technique, the concept seems odd, but the fact is, given the complexity of modern digital circuitry, it is easy to design systems which are incapable of being tested. It is even more common for an engineer to develop a set of test data for a product which tests all the conditions he considers important, but which cannot test many other conditions. The more complex a circuit becomes, and the fewer test points that are available, the more likely it is that testing will be incomplete or invalid.

The simplest type of fault simulator runs the simulated circuit using test patterns and expected outputs supplied by the engineer, and saves the resulting outputs. It then resimulates the circuit with a single fault inserted; that is, the simulator assumes that one point in the circuit is stuck at a logical zero or one and cannot be changed, representing one particular possible circuit malfunction. It records the output of this simulation, and compares the results to those of the fault-free circuit. If the two output patterns differ, the fault is said to be detected, since the specified test data will make this particular fault apparent if it occurs in the real circuit. The simulator then proceeds to resimulate the circuit with every possible fault of the type described, each time checking to see if the fault can be detected by the test patterns. Finally, it will usually report what percentage of the faults were detected, and perhaps list particular faults which were not detected.

The type of simulator described above is a serial fault simulator, so called because it simulates each fault in sequence, one at a time. A parallel fault simulator does essentially the same thing, except that it examines the fault-free circuit and several fault conditions simultaneously, using individual bits in the simulating computer's words to represent the logic states of several simulations in parallel. This technique is obviously much faster than the serial method, especially if the computer employed has a "wide" word; for instance, some simulators using supercomputers with a 64-bit word length can simulate 63 faults at once, and thus run the same simulation about four times as fast as an otherwise similar 16-bit machine.

A third type of fault simulator, called a concurrent simulator, is often the fastest of all. These programs utilize algorithms which examine a logic circuit's connections, and identify only those faults which are likely to be detectable by the simulation. By simulating, in effect, only those faults which potentially can be detected, the concurrent simulator reduces, sometimes drastically, the simulation time relative to other techniques. These systems require large amounts of memory for table-keeping and storage of intermediate results, and have become feasible only in recent years as the cost of memory has decreased. Concurrent fault simulators such as CADAT™ from HHB-Softron Inc. and HILO-3™ from GenRad Inc. are offered on many brands of engineering workstations today, taking advantage of the large amounts of memory typically available, and running in reasonable times without use of mainframes or supercomputers.

Engineering workstations can provide significant capabilities to the engineer who must develop comprehensive tests for complex circuits. After running a fault simulation, for instance, some workstations can highlight portions of a schematic diagram, showing those portions which were not exercised by the test program. Often the visualization made possible by this technique can assist the user in developing additional or different test procedures to improve the quality of the test. In addition, it is frequently possible to modify the circuit in a way that makes it more testable once the troublesome areas can be located. The entire area of testability and test pattern generation is being explored vigorously by engineering workstation manufacturers, software developers, and design engineers at this time. While there are no fully packaged solutions on the market today, there are many useful software tools available to the innovative engineer.

Daisy Systems Corporation offers a testability package consisting of a set of software that helps mechanize the iterative procedure an engineer uses with fault simulation to develop adequate test patterns. The Fault Grader runs a test pattern supplied by the user and scores it on the number of faults it detects. The user then can supplement the original pattern with additional tests designed to detect particular faults identified by the simulator. For each successive run, the grader predicts the incremental improvement attained. Engineers can use this set of tools to refine their test programs to whatever degree is necessary within their economic constraints.

Silvar-Lisco Corporation has described a system under development that allows their fault simulator to run each test pattern individually, followed by a graphical display of the schematic with detected and undetected faults highlighted (in different colors). This tool is intended to allow designers to assess difficult testing situations visually, exploiting the human's ability to make inferences and recognize patterns (tasks at which computers are still inferior to skilled engineers).

4.4 Analog Simulation

In the previous sections, we discussed logic and fault simulation tools that deal only with arbitrary states, whether a point in a circuit is fully on or off. By contrast, analog simulation is used to model the behavior of circuitry in a precise way, dealing in detail with voltages and currents as functions of time, frequency, temperature, and process variables. Most engineering workstations that offer this capability utilize variants of the SPICE analog simulator, initially developed at the University of California and refined by countless vendors and users over the last decade.

Analog simulators deal with precise mathematical models of circuit components. A passive element (resistor, capacitor, or inductor) is usually modeled with a single differential equation. An active component such as a transistor is modeled with a whole series of equations which relate the voltage, current, and time relationships among its terminals. Because of the nature and complexity of the circuit element models, even the solution of a simple circuit involves simultaneous solution of a sizable set of differential equations by numerical means. The demand of this application for computer power can be very high.

Since most workstations are based on rather modest computers, it is unusual to find a workstation capable of performing analog simulation in reasonable time for circuits much larger than a few hundred transistors. Indeed, most of the workstations based on personal computers cannot perform analog simulation at all; these systems generally produce an appropriate netlist for a particular simulator, which is then run on a larger system.

There are exceptions. Microsim Corporation offers PSPICE, a subset of the popular SPICE analog simulator, which runs on an Personal Computer. PSPICE provides all SPICE features except distortion analysis, and handles circuits up to about 120 transistors. The program's performance is said to be about 20% of a Digital Equipment Corporation VAX-11/780 super minicomputer.

Figure 4.4 illustrates the screen display of the CIRCUITPRO™

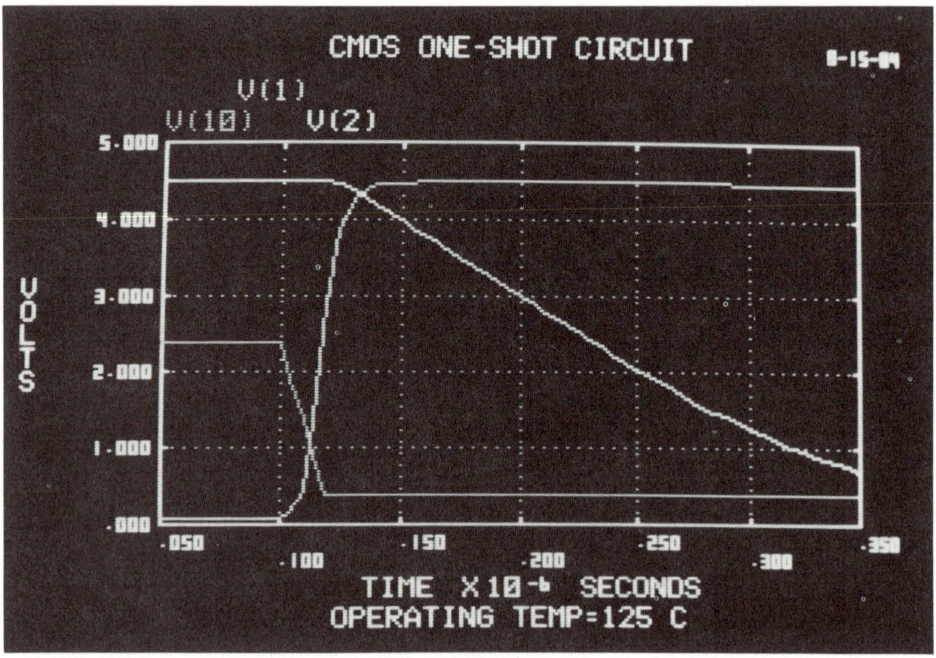

Figure 4.4 Many engineering workstations simulate the display of an oscilloscope. Here, the CIRCUITPRO™ package from E/Z CAD Inc. simulates the transient response of a circuit. (Courtesy of E/Z CAD Inc.)

analog simulation system from E/Z CAD Inc. This system also operates on the IBM Personal Computer and is said to be capable of handling circuits of up to about 150 transistors. The initial conditions are specified by the user, the monitor points are defined, and the simulator is run as a batch process. The resulting display, resembling an oscilloscope, illustrates clearly the relationships of three voltages, selected by the engineer, over a chosen time interval. Systems employing this type of display are available from several vendors on larger systems such as the Apollo and Sun Microsystems products, but are unusual on PCs.

Engineers contemplating design of analog circuitry, or the detailed design of integrated circuits, should carefully examine the simulation tools available on any workstations being considered for purchase, and make sure the system is capable of simulating circuits of the required size *on the workstation.* Overlooking this important capability issue may later necessitate use of a mainframe computer costing ten times as much as a workstation.

An innovative new system, the Analog Workbench™, was recently introduced by Analog Design Tools Inc. (Figure 4.5). This system is intended primarily for designers of high precision analog circuitry containing relatively small numbers of components, but requiring elaborate and extensive characterization. The screen display in the photograph shows the novel use of "icons" representing instruments found on the typical electronic bench: sweepers, voltmeters, and the like. The engineer connects these instruments to the schematic of his circuit and manipulates them as the circuit design is developed. This approach attempts to emulate the way engineers have worked in the past, rather than forcing adaptation to the ways of computers.

4.5 Physical Circuit Layout

The physical implementation of printed circuit boards and integrated circuits has been a major CAD/CAM application since the 1970s. Thus, it seems natural that this capability should appear also on engineering workstations. But layout tools have not been prominent in the first few years of the workstation era except in

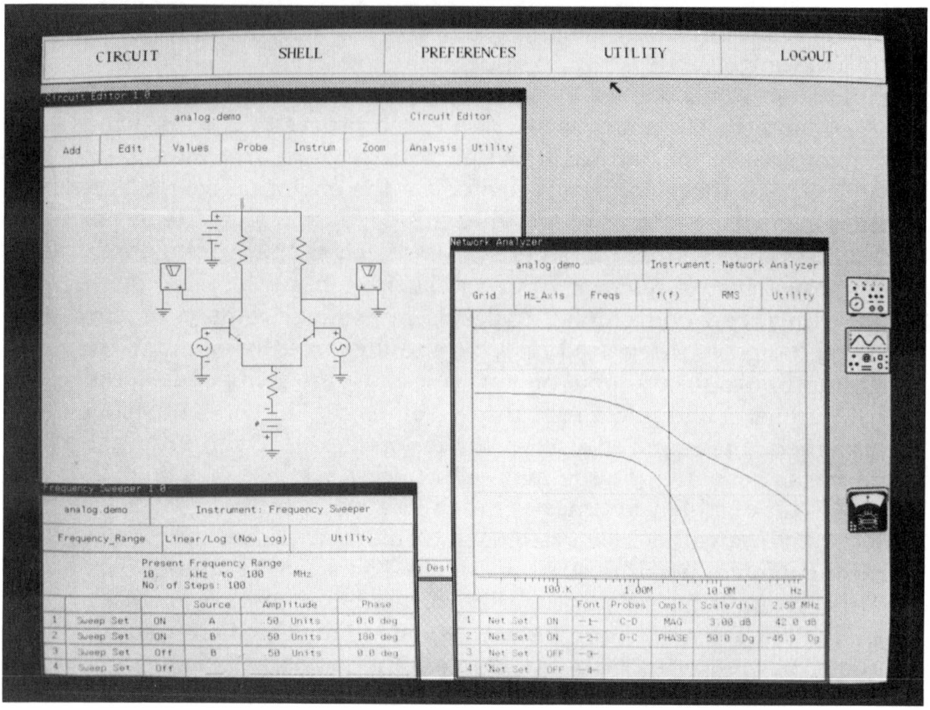

Figure 4.5 Icons representing instruments familiar to the engineer are used in the Analog Workbench, a workstation intended for use by designers of analog circuitry. The format of the display is intended to mimic the familiar environment of the electronic bench. (Courtesy Analog Design Tools Inc.)

one new application: semicustom integrated circuits, and particularly the type known as gate arrays.

4.5.1 Gate Array Development

Gate arrays consist of blocks of circuitry on standardized silicon chips, blocks treated as components by system designers. Designers specify to semiconductor manufacturers the desired connections between blocks, forming relatively complex integrated circuits in a way analogous to the wiring together of transistors and other parts in more conventional circuitry. The semiconductor manufacturer gains economy of scale, since the majority of the processing for every customer's circuit is identical; the user gains the ability to design a customized circuit function which replaces large amounts of conventional circuitry, saving power, space, and manufacturing costs. There are many other benefits to both parties of this technology: the semiconductor manufacturer saves on the cost and risk of product definition; customers obtain complex circuits which perform the exact function they need, rather than having to adapt a standard part to the application.

The physical layout and verification of gate array circuits is usually done by a specialized software package, frequently on an engineering workstation. Since semiconductor manufacturers design the detailed circuitry for each block in an array, users typically are concerned only with connections between blocks. Most gate array manufacturers now have cooperative arrangements with one or more engineering workstation vendors in which the manufacturer supplies physical and logical models of its products for use by designers. Workstation software combines the model information with the user's connection information to produce a complete set of data for the manufacturer to use in building the user's circuit.

A designer using gate arrays uses schematic editing tools, simulators, and physical layout tools in succession to proceed from a concept to a completed gate array design. Systems like the Daisy Gatemaster™ and the Mentor Graphics Array Station™ provide all the necessary tools for the entire process, from schematic through artwork generation. Many other systems provide some or all of the required capabilities, but are not marketed exclusively as gate array development systems.

Development of a gate array circuit in many ways is identical to development of any other digital circuit. Schematic editing, logic simulation, and fault simulation proceed in a fashion much like the descriptions earlier in this chapter. However, the user typically works with a development manual, supplied by the semiconductor manufacturer or the workstation vendor, which describes the peculiarities and constraints of the array for which he is designing. In a normal, fully customized design, no such guidebook exists, and the designer is free to implement anything he desires.

Gate arrays are physically implemented much like tiny printed circuit boards: metal connections between elements are formed using a photographic process, with the exact patterns being created by computer-driven artwork machines. The control information for these artwork machines (variously known as pattern generators, maskmaking machines, electron beam artwork systems, and many other names) is produced by software, which often runs directly on the designer's workstation. In other cases, the manufacturer prepares the actual artwork tapes, using as input a copy of the design engineer's data base from a workstation.

The generation of gate array artwork is usually done by a sequence of programs, running on the workstation, and controlled by the designer. The input to these programs is generally extracted directly from the schematic by a netlister. Next, a placer or placement program is run, which assigns the circuit elements specified in the schematic to specific, physical locations on the gate array. This process is usually automatic; it attempts to place elements in such a way as to minimize the amount of wiring necessary to implement the designer's circuit.

The next program is called a router because it finds routes for each connection between circuit elements. This process is usually automatic, but requires some user inspection (and sometimes modification) of the results. In most cases, a gate array router will be unable to make every single connection between elements automatically, because of constraints on the amount of physical space it has available for connections. For this reason, most modern routers have a postprocessing program which allows the user to modify and add connections to complete the circuit.

Routing postprocessors vary widely in sophistication. Some simply allow users to add and edit connections freely, a process that can allow serious errors to be introduced. At the other extreme are packages like the Mentor Graphics CADIGRAPH™ system, a restricted interactive graphics system incapable of allowing the user to create an error. The latter package displays all wiring the router was able to complete, and guides the user through the process of editing the remaining connections. A unique feature of this editor is that it will not allow wires to be placed if they are not connected as specified in the schematic, if they are too close together, or if they short-circuit to other connections. While the system does not guarantee that it will be possible to make all connections, it does guarantee that all connections made will be error-free.

In the event that the designer finds it impossible to make all connections in his circuit (not an uncommon occurrence), he must make one of several choices and iterate through the various software packages again. He may choose to modify or simplify the circuit, potentially making it easier for the software to do its job. He may elect to use a different array, one having more available circuit elements or different wiring constraints. Depending on the sophistication of the available software, he may even be able to perform some of the placement and routing manually, taking advantage of the human's superior ability to exploit pattern recognition. Packages like the Silvar-Lisco GRIP™ system allow great flexibility to the user in these areas, while always constraining the designer from creating errors.

The final step in designing a gate array circuit is to transfer the correctly placed-and-routed layout to the semiconductor manufacturer who will build the circuits. Each manufacturer has slightly different requirements for data entry, although most will accept a data base from any of the major workstation manufacturers. This is an area in which the guidelines are changing very rapidly; readers who are interested in exploiting gate array technology in their system designs should discuss their requirements in detail with each workstation manufacturer prior to purchase. Utmost caution is recommended!

4.5.2 Custom Integrated and Printed Circuit Layout

Design of detailed integrated circuit layouts is a technology that is almost exclusively the domain of semiconductor manufacturers. Indeed, this field is so specialized that few, if any, system designers have the skill or the inclination to attempt it. However, for that small group who have a need to perform detailed layout of integrated circuits, several of the major manufacturers of engineering workstations are currently developing or offering tools. Among the more prominent workstation-based packages are Daisy Systems Corporation's CHIPMASTER™ (Figure 4.6) and Metheus-CV Inc.'s Full Custom Design System (Figure 4.7). There are several other packages nearing introduction at the time this book is being written,

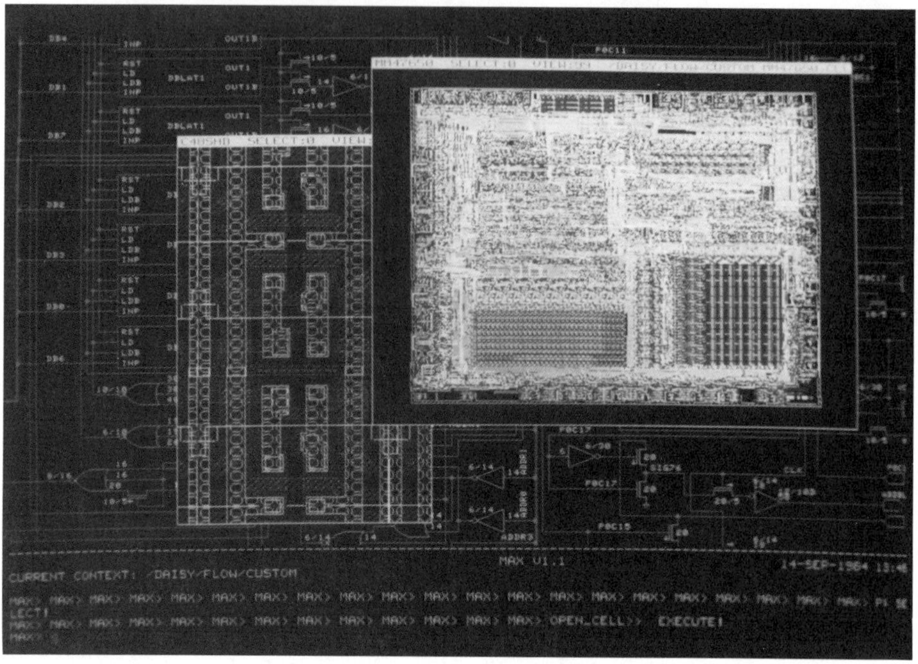

Figure 4.6 The CHIPMASTER^{TM} system is dedicated to the layout of complex integrated circuits. This display features an overall chip view, a detail view of a portion of the chip, and a view of the logic being implemented. (Courtesy Daisy Systems Inc.)

and the interested reader should have a large variety of capable alternatives to consider.

Likewise, the layout of printed circuit boards has traditionally been the province of a relatively small group of specialists. Recently, several prominent vendors have announced their intentions to market printed circuit design packages as part of their engineering workstation products. While comprehensive layout packages have been available for many years from a variety of vendors, the new packages, which operate in the workstation environment, should be expected to provide significant new ease of use and functionality.

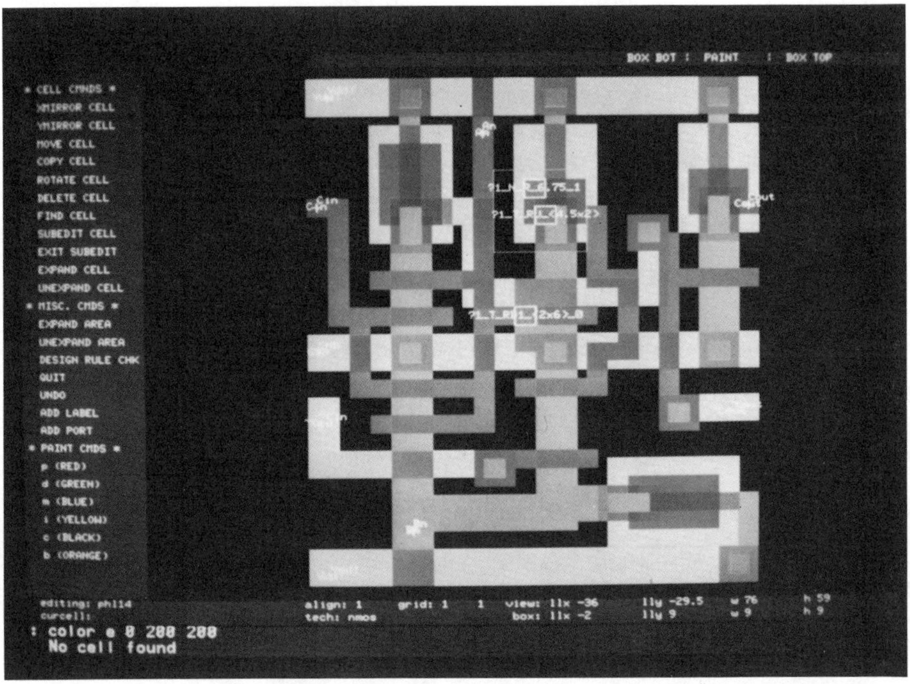

Figure 4.7 This detail view of a chip layout illustrates electronic parameters embedded in the physical symbols, indicated by the bright text. Access to both physical and electrical data is an important feature of many engineering workstations. (Courtesy Metheus-CV Inc.)

Printed circuit CAD traditionally has involved the transfer of hand sketched schematics from engineers to drafters, and further transfer of drawings to CAD system operators. In the new environment of interactive, workstation-based design, many manufacturers feel that the design engineer will want much more direct control of the board layout process. While the engineer's new tools are hierarchical, traditional printed circuit design tools are "flat" (they deal with entire circuits in terms of lowest level components, regardless of complexity).

Cadnetix Corporation, for instance, provides the CDX-9000

workstation for schematic entry and simulation, and the CDX-5000 for board layout. Both systems utilize similar data bases and are interconnected by a network, allowing very close coupling between circuit design and board design.

Placement and routing of printed circuit boards is conceptually similar to the process described above for gate arrays. However, on a board the designer does not deal with a fixed array of components that must be interconnected, but rather a "clean slate" upon which any type of components may be arranged in a variety of ways. Figure 4.8 illustrates the placement of components of various sizes on a board prior to routing. Figure 4.9 illustrates a completion editor in use, in which the designer is attempting to complete the routing of some connections that the software could

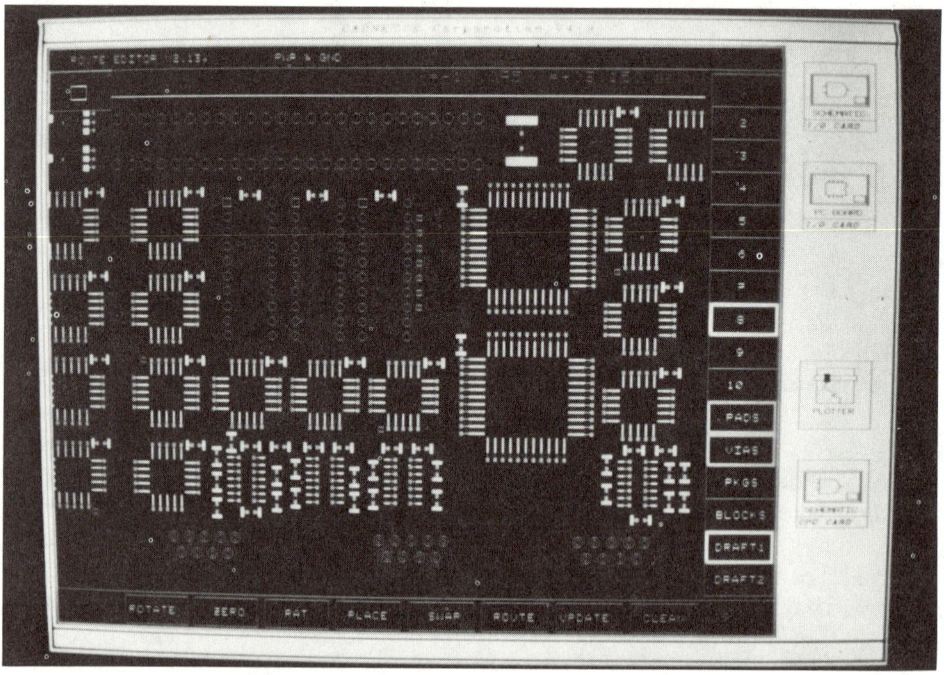

Figure 4.8 Initial placement of components on a printed circuit board is controlled by the designer on this Cadnetix workstation. (Courtesy Cadnetix Corporation)

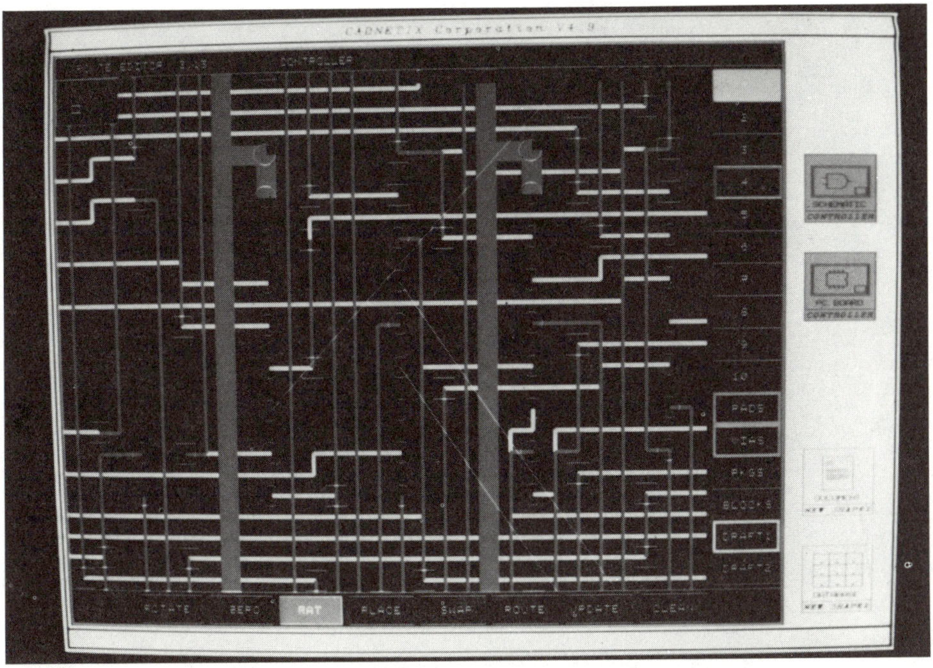

Figure 4.9 Routing programs rarely, if ever, are able to completely connect all paths on a board. A completion editor is used by the engineer to devise manual paths for the unroutes signals. (Courtesy Cadnetix Corporation)

not route (shown as faint diagonal lines). This sort of capability is likely to begin appearing on other vendors' systems in 1985, a process which will substantially broaden the potential marketplace for engineering workstations.

5

SOFTWARE FOR GENERAL ENGINEERING WORKSTATIONS

Although the electronics industry has been the most enthusiastic adopter of engineering workstations to date, there are many applications in other industries. This chapter will cover some of the software available in the mechanical engineering and architectural engineering fields today, as well as some general purpose applications found on workstations that have no counterparts on other types of systems.

Mechanical and architectural applications have not yet taken off in the same way that electronic applications have done, but most experts feel that major growth and innovation is soon to come. Dataquest Inc. [1] estimates that 71 percent of total CAD/CAM industry revenue in 1984 was for mechanical and for architecture, engineering, and construction (AEC) applications, and that only 23 percent was for electrical and electronic applications. These percentages are expected to change only slightly over the next five years, but the type of equipment sold will change radically: personal and standalone CAD/CAM equipment will grow from about 15 percent to nearly 50 percent of the market over the same period, reflecting the rapid growth anticipated in engineering workstations. Clearly, a great deal of growth is anticipated in workstation products for mechanical and AEC applications in the near future.

There are many reasons for the current emphasis on electronics by workstation manufacturers. Engineering workstations first appeared in computer and semiconductor companies as an answer to desperate, near term needs. The development of VLSI circuits (very large scale integration, an industry term for circuits consisting of from tens of thousands to perhaps millions of circuit elements), is a process so complex that CAE simply is not an option; to be a participant in this demanding business, a manufacturer must use computer aids, and use them quickly and efficiently. Since engineers designing VLSI circuits are strategically place to understand and exploit the latest in computer technology, it is not surprising that engineering workstation technology evolved in these companies and their suppliers first. In addition, the computing power required to process modern mechanical design analysis is so huge that economical hardware which could make full-function engineering workstations for mechanical design practical is not yet available. Traditional CAD/CAM systems for mechanical design applications have relied on mainframes and supercomputers for such computationally intense functions as finite element analysis and solid modeling image generation. As long as large computers remain a necessity, mechanical and architectural applications will lack the compelling economic factors that have driven the electronics industry to adopt engineering workstations.

The situation will clearly change when compact, economical computing resources with the power of today's mainframes are available. The operational advantages of distributed, personal computing and integrated tools are so strong that the industry is simply waiting for the economics of computing hardware to change.

Design tools for the electrical, mechanical, architectural, and electromechanical industries are well developed in traditional CAD/CAM systems. Most of the major vendors of CAD/CAM equipment have begun repackaging their existing systems in order to offer capabilities in these technologies on lower cost equipment. While these systems offer great capability and mature, powerful software, the integrating glue software that characterizes most electronic workstations (described in the previous chapter) is missing. For the most part, these repackaged systems are still not being optimized for use primarily by design engineers.

Even today, many engineers have had little personal experience with CAD tools, despite the fact that technicians and drafters in their companies have used them for years. Therefore, this chapter will review some of the major functions these systems can perform, with particular emphasis on the functions currently available on workstations.

5.1 Drafting Versus Modeling

Drafting is familiar to all engineers in all fields. Drawings are used to develop design concepts, to document products, to detail manufacturing requirements of a product, and sometimes to provide a realistic conceptual view of an object before it has been built. Despite the familiarity of drawings to all engineers, the nature of drafting on a board with pencil, paper, and straight edge is fundamentally different than on a CAD system: these systems usually distinguish between a model and a drawing, concepts that should be clearly understood by engineers who intend to use them.

The user of a CAD system creates a three-dimensional geometric representation of a part by using commands to develop lines, arcs, surfaces, conics, etc. The system enters into its database the three-dimensional coordinates of these entities as sets of numbers with descriptors that define what type of object they represent. The display that the designer sees on the screen of his workstation is of course two dimensional, but it is only a flat projection of the three-dimensional data stored in the system. Thus, it is usually possible to look at the model from any desired direction, and at any desired scale, without disturbing the database at all; the user defines the three-dimensional coordinates of his desired viewpoint and window, and the system performs the necessary mathematical transformations to produce the display.

Drawings are in many ways like screen displays: they are

two-dimensional representations of three-dimensional objects, produced at various scales and viewpoints. In a sophisticated CAD system, a drawing will only reference the geometric model of an object; thus, the drawing outline, dimensions, annotation, text, etc. that must be present on a useful drawing is stored separately from the model in the database. Each of the many views of an object can be described and annotated, all referring to a single model. If the model is changed by using the system's editing capabilities, all drawings that reference it will automatically reflect the change, in all views. This extremely powerful capability exists in most full-function CAD/CAM systems used for mechanical design, such as the Computervision Designer™ series and the Applicon BRAVO!™.

5.2 Drafting Systems on Workstations

Most drafting packages used on traditional CAD/CAM systems deal with portrayals of three-dimensional objects. This requirement means that these software packages must deal with much more complex mathematics than two-dimensional packages, and thus require substantially more computer resources. However, many drafting systems on small machines, particularly on personal computers, are simple two-dimensional packages. These systems lack the concept of an underlying model, and are only capable of creating lines, arcs, and text on a flat drawing. This fundamental difference between a design system and a drafting system must be recognized when evaluating low-cost workstations: there are good reasons for the low cost!

One should not interpret these facts to mean that two-dimensional systems are not useful though. AUTOCAD™, a product of Autodesk Inc., may well be the most popular and widespread CAD package available today, despite the fact that it is capable only of two-dimensional drafting functions. This system is available on IBM Personal Computers as well as most compatible systems. It is useful for a wide range of applications, including architectural drawings, design of flat patterns, floor layouts, flow diagrams, piping schematics, and many other tasks.

Another very popular two-dimensional drawing system is the design oriented graphics system, or DOGS™, produced by PAFEC

Engineering Consultants Inc. DOGS is available on a wide range of computers, ranging from personal computers to standalone workstations like the Apollo family to mainframes like the Data General Corporation MV-10000.

Some three-dimensional drafting packages with significant and useful capabilities are available on personal computers, and many are available on more powerful workstations such as the Apollo and Sun families. The recent introduction of the IBM PC/AT, a more powerful version of the original PC, has resulted in the rapid appearance of sophisticated yet economical software for drafting. For instance, Manufacturing and Consulting Services, Inc. (MCSI) has recently introduced Anvil™ 1000 MD, a scaled-down, two-dimensional, yet compatible version of its popular mainframe-based Anvil 4000 design and drafting system.

5.3 Repackaged Drafting Systems

Many established vendors of CAD/CAM software are beginning to reconfigure their popular software systems to run on the compact, powerful workstations produced by Apollo, Sun Microsystems, Digital Equipment, Data General, Ridge Computer, Masscomp, and many other companies. For instance, Applicon Inc. has introduced the ARIA™ workstation, a low-cost version of their BRAVO! system based on a small Digital Equipment Corporation VAX™ computer, capable of running most of the vendor's existing CAD/CAM packages for mechanical design, drafting, plant layout, etc.

Computervision Corporation is marketing its CADDS™ system on the IBM PC/AT. This system runs a version of Computervision's existing software for mechanical and architectural design and is compatible with the vendor's larger minicomputer-based systems.

Auto-Trol Corporation now offers its popular Series 5000 and 7000 mainframe software on the company's Advanced Graphics Workstation, a customized system based on Apollo computers of various models. Likewise, Calma Company now offers much of its traditional CAD/CAM software on Apollo computers.

In summary, most of the larger, traditional system vendors have recognized the trend toward personalized use of CAD/CAM and CAE software, and have made an initial response. Their approach, repackaging existing systems, is understandable when their enormous investment in software is considered. In addition, this evolutionary approach is likely to ease existing customers' transition to workstation technology.

5.4 Finite Element Software

Finite element analysis is perhaps the most difficult CAD function to implement on engineering workstations because of its appetite for computer power. Since the concepts may not be familiar to all engineers, the following brief overview of finite element analysis will provide some understanding of the problem. In keeping with the goals of this book, needless detail will be avoided. Literally hundreds of books and papers have been written on the subject at all levels of detail; interested readers are encouraged to explore the subject in depth in one of the many specialized works [2-4].

Finite element analysis is a technique for simulating the behavior of a variety of systems by subdividing complex assemblies into a large number of mathematically simple elements. The most common use for the technique is in analysis of stress, vibration, and heat flow, although there are many other uses of the same mathematical principles in other fields.

The analysis of stresses in a simple two-dimensional part is easily illustrated and understood: the part is divided into a mesh of small rectangular and triangular elements in such a way that the elements exactly fill the area of the part. Curved edges and discontinuities are approximated by using a large number of elements, while broad, uniform sections of the part can be filled with a few large elements. Each element is encoded for computer analysis in terms of the coordinates of its vertices and its connections to adjacent elements (such common edges are referred to as nodes).

Following the generation and encoding of a mesh, the user encodes the location, magnitude, and direction of the stress vectors to be applied to the part, and the location of any constraints. The analysis program then constructs a set of simultaneous equations which model the behavior of each element. This type of analysis assumes that each individual element is bound to its neighbors, and can be described by a simple linear equation relating the imposed stresses to the resulting strain or deformation. The program performs a simultaneous solution of all the element equations, resulting in a matrix of the deformation values associated with each element.

From this simplified conceptual description of the technique, several facts should be obvious. First, even the simplest parts will require a large number of elements for accurate modeling; for instance, if a small number of straight-edged elements are used to model a curved edge, it is fairly obvious that the model will only approximate the actual behavior of the part. Second, most problems do not consist of simple, plane parts. In fact, most finite element analysis problems actually deal with three-dimensional structures and use solid elements (tetrahedra and hexahedra) rather than plane elements. Obviously, this greatly increases both the number of equations to be evaluated and the complexity of the equations. Third, the accuracy of the analysis cannot be any better than the generation of the mesh. Errors in encoding the mesh, or in constructing it, or poor choice of element type or density, can invalidate an analysis. CAD systems provide a great deal of help with these problems, and have been instrumental in the widespread use of finite element analysis techniques in many industries.

5.4.1 Mesh Generation

Mesh generation requires a skilled analyst who understands the limitations and behavior of finite element analysis software. Nevertheless, the interactive graphics capabilities of a CAD system are invaluable to the analyst. There are several software packages that can generate meshes automatically, given some guidance from the analyst. The analyst typically assists the software in resolving ambiguities in the model (such as the difference between surfaces and holes), supplying suggested location and density of node points, and making manual corrections in places where the software has made mistakes (such as placing elements outside the actual boundaries of the part).

Mesh generation software is available on sophisticated turnkey CAD/CAM systems, and many of the important vendors are following the established pattern of "porting" these applications to advanced engineering workstations. For instance, the Graphic Finite Element Modeling package offered by McDonnell Douglas on its Unigraphics™ system is now available not only on the vendor's traditional mainframe systems, but also on Data General workstations. MacNeal Schwendler and MCSI both offer versions of their interactive mesh generation software on IBM personal computers in addition to larger systems.

5.4.2 *Postprocessing*

Another important function of finite element systems is to transform the enormous amount of data produced by the program into a meaningful representation for the analyst. The most effective means of doing this is to use a graphics display. For instance, a shaded picture of the mesh can be displayed, and a color assigned to each element related to its stress, deformation, or temperature. Typically, "cool" colors like blue are assigned to low-stress regions, while "hot" colors like red are assigned to high stress regions; the resulting pictures dramatically illustrate stress concentrations as well as modeling errors. The analyst can use the interactive graphics routines of a workstation to examine critical regions, to refine or correct meshes, and to even modify the model in order to correct design problems.

Postprocessing functions are well suited to engineering workstations in that they require high performance graphics, personal access, integration of several software modules, and only modest computational performance. Packages such as PATRAN™ from PDA Associates Inc., ANVIL™ from MCSI, MSC/NASTRAN™ from MacNeal Schwendler, and Unigraphics™ from McDonnell Douglas are increasingly available on a variety of workstations from the major vendors.

5.4.3 *The Analysis Problem*

The actual analysis programs used in finite element analysis are very large, and require huge amounts of file space. It is common for runs to last for hours on large mainframes, and consume hundreds of megabytes of disc space for their work files. Programs such as MSC/NASTRAN™ from MacNeal Schwendler Corporation and ANSYS™ from Swanson Analysis Systems Inc. are currently offered by the vendors on the Cray-1 supercomputer, and super minicomputers costing $500,000 like the Digital Equipment Corporation VAX™, the Prime Computer 9950, the Data General Corporation MV-10000, and the Harris 1000. Such applications are clearly beyond the capabilities of even the most powerful engineering workstations today, and will remain so for years to come [5].

The systems advertised today that claim to perform finite element analysis on IBM PCs and Apollo computers do work, and can perform useful analyses, but only on very restricted problems. For instance, MacNeal-Schwendler Corporation's MSC/pal™, a two- and three-dimensional finite element analysis package for use on IBM personal computers is said to analyze mechanical systems composed of beams, plates, springs, dampers, and many other elements in static, normal mode, transient and frequency response. However, the analysis capabilities are severely limited by the capabilities of the PC. This package is useful for training and education in analysis techniques, and, since the input data for MSC/pal can be converted for use with the vendor's MSC/NASTRAN™, a much larger package for use on mainframe computers, it is useful as a low-cost front-end system. This compatibility between limited- and full-capability tools represents an innovative approach taken by the vendor to explore the potential of mechanical engineering workstations, an approach sure to be followed by others in the near future.

5.5 Architecture, Engineering, and Construction

Applications for architecture, engineering, and construction are well suited to engineering workstations in most cases. These applications typically are specialized cases of two- and three-dimensional drafting whose end product is the drawing itself (Figure 5.1).

Architecture, in particular, represents a fruitful field for CAD techniques and for workstation companies. For instance, high quality renderings can be prepared using interactive graphics systems and can be customized and artistically embellished far easier than can manual drawings. Detailed, multiview plans can be produced and updated far more easily on a graphics system than manually. Finally, the opportunity exists (although it is still not a common practice) to transfer drawings electronically to subcontractors and design firms using discs and tapes, instead of the traditional rolls of hand-drawn blueprints. This technique not only is easier, but it assures absolutely accurate transfer of information.

A popular architectural system based on a minicomputer, the

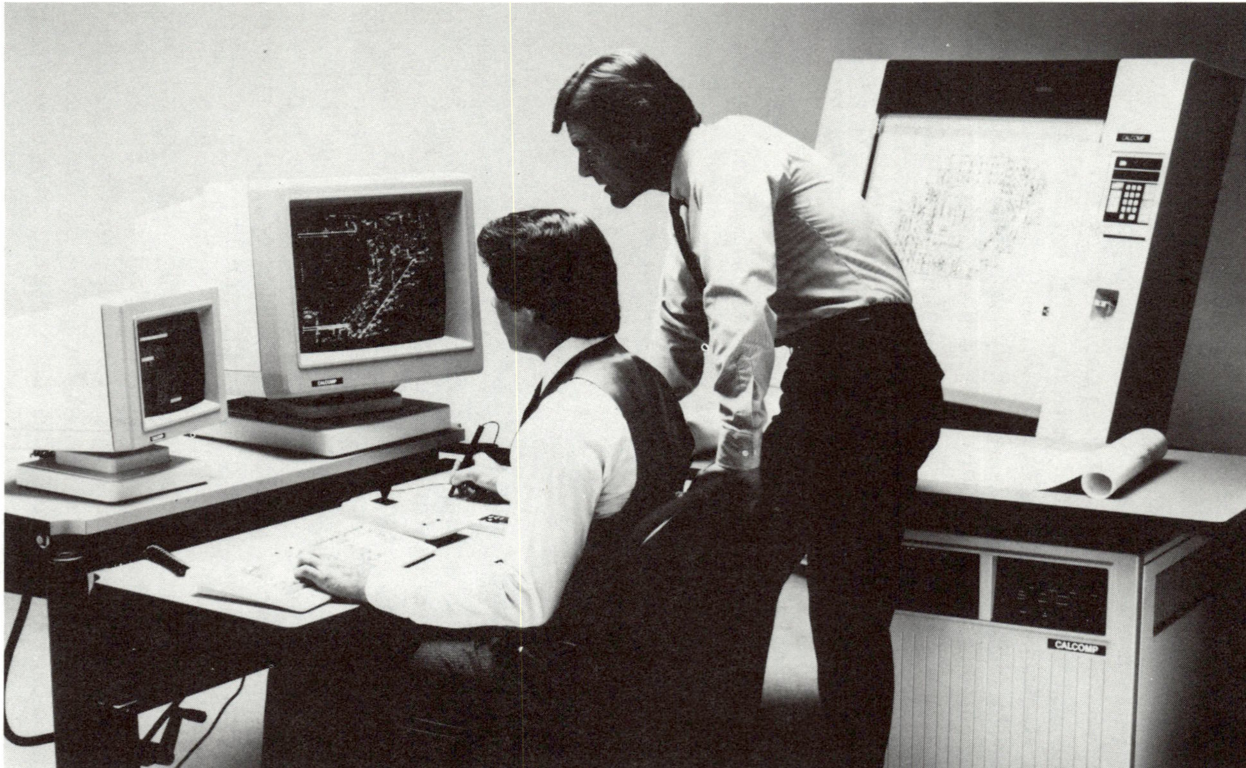

Figure 5.1 Architectural applications for engineering workstations produce high quality drawings as their primary output. This system employs a high speed four-pen plotter. (Courtesy Calcomp, a Sanders Company)

Figure 5.2 The Calcomp System 25 employs a high speed Picture Processor to allow panning and zooming in real time throughout complex architectural drawings. (Courtesy Calcomp, a Sanders Company)

Calcomp IDS, has been repackaged as a high performance workstation based on a Masscomp computer and a high performance graphics accelerator called the Picture Processor (Figure 5.2). The System 25 provides many functions, in addition to drafting, useful to architects. For instance, it provides a very elaborate stacking and blocking package used to study and optimize the allocation of space within multiple, multistory buildings. System 25 uses the UNIX™ (AT&T Corporation) operating system and the Ethernet™ (Xerox Corporation) local area network to interconnect large clusters of workstations, and to permit easy access to larger computers. This distributed architecture allows many architects and engineers to participate in a design, and to share databases.

Holguin & Associates offer the CEADS™ series of systems based on Hewlett-Packard and Wang small computers. The CEADS-CADD package is a general purpose two-dimensional drafting package aimed at architects and civil engineers. In addition to the usual drafting primitives like lines and arcs, this package offers several big-system features like primitives for doors and windows; when the designer moves one of these elements, the system automatically provides the correct wall terminations. The CEADS-CIVIL package includes several capabilities tailored to civil engineering applications, such as earthwork cut and fill calculations.

BruningCAD offers the SPECTRA system (based on the Hewlett-Packard series 9000 workstations) with its EasyDraf-2 architectural software. This package features "smart" walls which automatically open and terminate with window and door movements. A unique feature is AttriBase™, a spreadsheet-like capability that incorporates bill of materials processing. Items changed in drawings are automatically updated in the user's spreadsheets, keeping material requirements up to date continuously.

Computervision Corporation's SpacePlan/3000™ runs on the company's series 3000 systems (based on the Sun-2 workstation from Sun Microsystems). The system provides tools for analyzing organizational relationships within a building and generates block layouts with alternatives; the user can evaluate alternatives and apply subjective criteria to reach the best arrangement of space, rather than accept a singular computer-optimized layout.

Many of the smaller two-dimensional drafting systems are also quite useful. PC based systems such as CAD Master™ from Datagraphic Systems, CADPlan™ from Personal CAD Systems Inc., and AutoCAD™ from Autodesk Inc. are commonly used to prepare site plans and office layouts.

5.6 Technical Publications Software

Engineering workstations have made possible the development of specialized software for the creation of technical publications, manuals, and data sheets. While this is not strictly an engineering function, it is illustrative of the kind of innovative solutions that can be brought to bear on problems when low cost graphics, unified databases, and distributed processing become common in engineering organizations.

Traditional technical publications packages have relied heavily on the tools of the publishing industry and specialized, costly dedicated computer systems. Tools available to the engineering department have typically consisted only of text formatters and word processors. Until very recently, it has been impossible to intermix text and graphics in a document without resorting to customized programming or manual cut-and-paste methods.

Some engineering workstations now provide very elaborate document processing tools. Perhaps the most highly developed is the package simply called DOC™, a product of Mentor Graphics Corporation. DOC begins with the usual functions of a word processor: it provides a very powerful full screen text editor with commands for searching, substitution, justification, margin control, headers and footers, and so on. It then adds the ability to control the character font used for printing, using the extensive graphics capabilities of the Apollo computer. DOC also allows the user to incorporate graphics into documents *by reference.* For instance, if it is desirable to incorporate a portion of a schematic diagram in a manual, the user sets up a description of the location of the schematic database, the graphical "window" of the schematic to be displayed in the document, the scale and rotation at which it is to be portrayed, and so on. The resulting view then appears on the screen exactly as it will look in the final document.

Such figures within documents have many interesting and useful properties. Since a figure is only referenced and not included in the document database, the figures in the document reflect the exact current version of the graphics database, including all changes and edits, at the time of printing. Thus, the figures in a manual are automatically updated when a schematic is changed, without any effort on the part of the user. If new figures and text are added to the document at a later date, the DOC package automatically adjusts the position of figures on the appropriate pages, adjusts figure numbers, and even adjusts the page numbers in figure references on other pages, both forward and backward.

Another very elaborate documentation system is TechPubs/3000 from Computervision Corporation. This package runs on the

CV 3000 line of workstations, and can incorporate graphics images from other software packages in documents in much the same fashion as Mentor's DOC. While the Mentor system is very elaborate, it is useful only with the vendor's electronics database and pictures hand-drawn with the PICED™ two-dimensional drawing package. TechPubs/3000, on the other hand, works with graphics images from CV's Schematic Capture/3000, Logic Design and Simulation/3000, Drafting/3000, Spaceplan™/3000 and Factory-Vision™/3000, making it applicable to a wide range of disciplines.

5.7 Final Notes

In this chapter, we have briefly covered some of the engineering workstation products currently available for applications other than electronics. Although there are many products, in most respects they have few capabilities beyond those of previous CAD/CAM systems. Frequently, because of the modest computers they use, they are less capable than their predecessors.

Readers should not interpret the lack of radically new products as a sign that these industries are being overlooked by engineering workstation manufacturers. As stated previously, the major impediment to the widespread use of low cost hardware and integrated systems is the unavailability of very high performance, compact, economical computers. This situation, as any reader of technical publications is aware, is changing daily. By the end of the current decade, it is likely that workstations with the power of mainframes and the cost of personal computers will be available. When this happens, dramatic new capabilities will be implemented for mechanical design and architecture, engineering, and construction applications.

References

1. Focus Conference, Dataquest, Inc., "Workstations: The Evolution of an Industry," September 21, 1984, Boston, MA.
2. C. Hayden Hamilton III, CAD/CAM Techniques for Generating Finite-Element Meshes, in Computer-Aided Design, Engineering, and Drafting, Auerbach Publishers Inc., Pennsauken, NJ, 1985, Article 2.2.2
3. L. A. Schmit, Structural Analysis—Its Genesis and Development, Journal of the AIAA, volume 19, number 10, October, 1981, pp. 1249-1263.

4. B. Fredricksson and B. G. A. Persson, Finite Element Programs in Integrated Software for Structural Mechanics and CAD, *Computer Aided Design*, volume 13, number 1, January, 1981, pp. 27-39.
5. Henry H. Fong, An Evaluation of Eight U.S. General Purpose Finite-Element Computer Programs. Paper presented at the 23rd AIAA/ASME/AHS Structures, Structural Dynamics, and Materials Conference, May, 10-12, New Orleans, LA.

6
INTERFACES

No CAD/CAM system stands entirely alone. No matter how good or how comprehensive a system may be, it is always necessary to access the information created on the system and use it elsewhere. Thus, any study leading to the purchase of engineering workstations should include an analysis of the interfaces needed to make the equipment productive in your environment.

6.1 Interfaces to Other Tools

The simplest interfaces are those that extract information from the database of the workstation and reformat it to drive some external tool. For instance, if your design facility already owns an expensive pen plotter, the vendor may provide a software package which will allow the use of that plotter by users of the workstations. If numerically controlled machine tools are already installed, the vendor may provide a means for creating control tapes directly from the database.

Potential buyers of engineering workstations should be very cautious in purchasing interface software from vendors. Traditional suppliers of CAD/CAM systems have invested literally thousands of man-years of software development effort in their NC programming software; it is unlikely that any of the startup companies supplying workstations will be able to provide such comprehensive capabilities for several years. Indeed, those which supply low-end workstations, perhaps using a personal computer as the host, may not supply any such capability.

Interfaces are critical to any purchase of workstations. Before buying systems from any vendor, decide exactly what external tools and equipment must be driven from the workstations, and insist on a realistic demonstration of each interface, using your own data and your own external systems. This effort is much more difficult than it sounds; it may require learning enough about the systems *prior to purchase* to create a design representative of your intended use of the system. Remember, simple demonstrations are not likely to prove whether an interface is complete. You must demonstrate *all* the capabilities you require, or you are unlikely to discover any shortcomings until it is too late.

6.2 Interfaces Between Systems

There are few standards for databases in the CAD/CAM industry, and none of them are sufficiently comprehensive to guarantee that a design can be transferred intact from one type of system to another in all cases. This simple fact is crucial to buyers of any type of CAD/CAM system because it means that, with only a few exceptions, you will be committed to, and dependent on, one vendor for a crucial part of your business. Particularly with engineering workstations, the vendor may be a very immature company which is growing at an astonishing rate; the all too common result of this situation is poor product support. Less common, but even more serious, is the prospect of the vendor disappearing altogether. Every purchaser of these systems should at least consider the consequences of a change of vendor after a considerable investment in database has been made.

Purchasers of engineering workstations are in many ways pioneers. The risks of purchase can be high, but the rewards may well be worth the risk. Dealing with the risk is usually a matter of protecting one's database: if the vendor goes bankrupt, or fails to support the product, or reneges on commitments, how does the buyer continue to do business? How can the data, the libraries, the training, be transferred to an alternative system? These are the same questions which have been asked for years by users of CAD/CAM systems: some of the answers, and many of the pitfalls, while well known, will be described below.

6.2.1 Direct Transfer

To transfer data from one system to another, it is possible (at least in principle) to write a software package that translates each item of data from the format of System A directly to the format of System B. Writing direct interface programs between systems requires that the writer understand the exact, detailed representation of data used by each vendor. Unfortunately, this information is usually not published, and is even considered to be confidential by many vendors. Thus, it is usually very difficult to write a good translator between any two given systems.

A second problem with this type of translation is that vendors can, and frequently do, change the format of their internal data without notice to customers. Such changes are usually part of major software releases, and are accompanied by many desirable new features and "bug fixes." The system's users generally want to install new software releases immediately because of the added features and fixes. However, if a custom translator has been in use, it is frequently found that the translator no longer will work when the new software release is installed, and that there is no documentation of internal database changes. As a result, the translator may not be functional for some time, causing disruption of the flow of work and lost production.

As if the problem of database format were not enough, there is another problem that usually serves to make direct translators impractical: to interchange between two systems requires at least two programs (from A to B, and from B to A). For three systems, eight programs are required, four systems require sixteen programs, and so on. Clearly, the direct translation between systems is impractical in most cases, and, as a result, there is little hope of obtaining such software unless you write it yourself. This is almost never a good idea, unless your company has a very competent software staff prepared to support the program for many years.

6.2.2 *Indirect Transfer*

An indirect transfer solves many problems in converting data from one type of system to another. Rather than utilize a single program that reads from A and writes to B, this method employs two separate programs and a neutral, intermediate format. Thus a program now reads the format of System A and writes the intermediate format, and a second program reads the intermediate format, writing its output to System B.

An immediate advantage of this method is that each program only requires intimate knowledge of one system's data format; thus, it is practical for system vendors to write and support such programs, and to incorporate any database changes in each release of software to customers. A second advantage is that there need only be two programs, a writer and a reader, for each manufacturer of systems, rather than "two to the 'n' power" programs. The major problem of indirect interchanges is the difficulty of establishing the intermediate format in such a way that it satisfies all users and supports the features of all systems. Nevertheless, there have been a few moderately successful attempts to define interchange standards.

6.2.3 *Initial Graphics Exchange Specification*

Initial Graphics Exchange Specification (IGES), an interchange format specifically designed to relieve users of mechanical design systems of the vendor-dependence problem, was originally developed under the United States Air Force's ICAM (Integrated Computer-Aided Manufacturing) program with coordination by the National Bureau of Standards. A task team led by the National Bureau of Standards, and consisting of representatives from a number of large government contractors, published the initial specification for IGES in 1980. Since then, the format has been augmented several times through a committee of the American National Standards Institute, and is currently supported by at least 20 vendors of CAD/CAM systems [1].

The IGES Version 1.0 specification supports only basic three-dimensional wireframe geometry, text, and dimensional data. Thus, it is at best a means for archiving and transferring drawings. Version 2.0 contains definitions for finite element modeling entities, some advanced curves and surfaces, and some electrical product data. Future versions of the specification will include specific data for a number of specialized fields, such as piping, solid models, and mapping. Eventually, the proponents of IGES say, the standard will be robust enough to support true interchange of designs, and not just drawings.

In the mechanical engineering and drafting industries, IGES has been of significant value in reducing the risks of vendor dependence. Hopever, a true standard like IGES seems always to fall short of being an ideal solution to the problem it addresses. A major difficulty is that it inevitably must represent a compromise between many companies and approaches. Thus, it cannot provide a means for transporting every nuance, every feature of System A's data base to System B. A second problem, common to any committee activity, is slowness. From the time a need for a new feature is generally recognized until it becomes a part of the published IGES standard, two years typically elapse. Today, as solid modeling is becoming widespread, there is no standard for transporting solids data from one system to another. Reliable software running on several vendors' systems may not be available until at least 1987. In the meantime, system users are vulnerable to dependence on a single vendor.

Today, engineering workstations for mechanical design are only beginning to emerge in commercial form; since an accepted industry standard already exists which promises to someday make possible the interchange of design information between systems, purchasers of these systems may well find that they will have IGES capability from the outset. This happy situation may well be unprecedented in the CAD/CAM industry, and should make the transition from existing CAD/CAM systems to engineering workstations much smoother for the mechanical design industry than it has been for the electronics industry.

However, the reassurance is incomplete: IGES does a good job of interchanging drawings and dimensional data between systems, but is weak in properties information. For instance, each CAD/CAM system for mechanical design allows the attachment of nongraphical information to objects in the drawing, such as weight, material, part number, etc. IGES does not provide a universal method for transmitting these data from one vendor's system to another; indeed, these are the very types of data that distinguish an engineering workstation from a drafting system.

6.2.4 Electronics Industry Standards

In the electronics industry, the situation is more difficult than in the mechanical design industry. Users of electronic design CAD/CAM systems have no accepted standard yet for interchanging data between different vendors. In many cases, it is even impossible to protect against changes in a vendor's *own* data base. Users of some systems have found that the data they carefully archived in June are useless in December because their vendor has changed the internal representation of data. Despite the risks, the standards in electronics CAD/CAM have typically been certain vendor formats adopted by other vendors out of competitive necessity.

Calma Company's GDS-II Stream Format is one such de facto standard extensively used in the electronics industry, particularly for integrated circuit and printed circuit applications. This format was originally intended by the vendor to be a means for users to save archive data without concern over database changes; Calma promised to maintain the Stream Format regardless of internal changes. Stream Format provides a means to describe geometric shapes, subfigure references, text, and user-defined properties. Since Calma's equipment is used in a very large percentage of all electronics firms, the format quickly became a convenient means for users to exchange data with each other and with contractors. It has been adopted by many service bureaus as a standard for accepting data to be plotted or edited, and is also supported by many of Calma's competitors on their own systems.

De facto standards like Stream Format are useful, but limited; if the originating vendor chooses to change or augment the format (and there are no guarantees in most cases), other vendors are forced to follow. Frequently, the process of following is a matter of trial and error, depending on the original vendor's willingness to share format details with competitors and users. A true standard may only be changed through a mutual agreement between parties to the standard, and provides a much more reliable, if slower, means of avoiding vendor dependence.

6.2.5 Electronic Design Interchange Format

The Electronic Design Interchange Format (EDIF) committee was formed in early 1984 in an attempt to create a standard for workstation users and vendors quickly, before the industry settled on the sort of pseudo-standards used by the electronics CAD/CAM industry. EDIF would not only provide for interchange between systems (a goal supported more by customers than by vendors!), but would also provide a means for vendors of semiconductor devices to support libraries of their products on the dozens of different workstations being used by their customers [2]. The committee is an unusual effort, in that it is not sponsored by the government or any traditional standards organization: it is a cooperative effort between a group of electronics companies (Motorola, National Semiconductor, Texas Instruments, Tektronix Inc.), two prominent workstation vendors (Mentor Graphics, Daisy Systems), and a university (University of California at Berkeley). The membership of the committee was deliberately restricted to minimize the two major problems with standards activities cited in the previous section; slowness to act, and technical compromise. It was felt that, if the standard could be defined rapidly by a small group of highly knowledgeable people representing only three interest groups, that agreement could be reached before the industry had developed ad hoc standards. The initial release of the specification was published in November, 1984 and has been enthusiastically supported by workstations users and semiconductor vendors.

The EDIF specification provides a comprehensive framework for all the types of information present in a complete electronic system design. The format is open-ended; that is, it does not attempt to limit the types of information that may be represented. Rather, it provides a standardized way to represent new data types as they are devised, and a means for users and programs to skip over data that are not understood or needed for any particular application.

The modular, open-ended structure of EDIF has allowed much of its development to proceed in parallel; the seven institutions working on its development have been able to subdivide the problem and work on their own pieces somewhat independently, with confidence that everything will fit together in the end (note the resemblance to the methodology used so successfully by workstation users).

EDIF from its inception was designed to encompass all aspects of electronic design. IGES, by contrast, was originally very limited in scope, and has added more sophicticated capabilities gradually. The intent of EDIF is to define the framework in advance, so that manufacturers of workstations may utilize it as their systems evolve. The EDIF committee has publicly stated that they do not intend to become a new standards committee: they intend to transfer the standard to ANSI or IEEE as soon as its major features have been defined and accepted by users. It now remains to be seen whether, and how quickly, the standard will be adopted by the major workstation vendors.

References

1. IGES Assessment: Present and Future, Curtis H. Parks. In *CAD/CAM: Management Strategies*, Auerbach Publishers Inc., 1984.
2. Preliminary EDIF report, version 1.0.0. The EDIF User's Group Design Automation Dept., Texas Instruments, P.O. Box 225474, MS3668, Dallas, TX 75265.

7
ACQUISITION OF ENGINEERING WORKSTATIONS

Assuming that you have decided that your company should acquire engineering workstations, there are a number of important questions to be answered before the systems can be purchased. This chapter will cover the major steps to be taken to assure that you find the right workstation vendor, and help you to prepare a thorough justification. Many books and articles have been written on the subject of justification of capital equipment; only the considerations special to engineering workstations will be covered here, since it is assumed that the reader is familiar with the general principles.

CAD/CAM systems are typically justified using time study methods, measuring the increased productivity expected in various activities. However, engineering workstations may require a different approach. In Chapter 1, the essential characteristic that makes a workstation system different from a CAD/CAM system was described: the former is used predominantly in the creative parts of product development, while the latter is typically used for drafting and documentation. The efficiency of drafting can usually be measured with some accuracy, but conceptual design is inherently difficult to measure. Thus, a different strategy to measure the impact of workstations on organizations may be needed. One such method is discussed later in this chapter.

7.1 Vendor Selection Cautions

There are dozens of vendors of engineering workstations, ranging from tiny "garage operations" to familiar CAD/CAM vendors. However, the rapid growth of the industry precludes listing them all in a book of this type: lead times are so long that many of the vendors listed today will be gone before the book is in print. Likewise, the history of the industry shows that there will be new companies in a year which do not exist now, many of whom may well be the new technology leaders. Therefore, this chapter will give only an overview of the major vendors of 1984, and some characteristics common to most of today's companies.

One characteristic common to most workstation vendors is phenomenal growth. A typical success story is illustrated in Figure 7.1. Apollo Computer manufactures workstation hardware and operating systems software which has been adopted by a large variety of engineering workstation vendors. Apollo is by no means unique; companies such as Sun Microsystems, Masscomp Inc., and Ridge Computer Corporation all supply systems to workstation vendors, and all are growing exponentially as systems companies build workstations based on their products.

With the enormous growth rates of such companies can come major problems for customers. Frequently in new companies experiencing rapid growth, there are problems with customer support, field service, and quality control. When you select a vendor for your engineering workstations, be sure to look beyond their obvious recent successes and predictions of future growth. If the company is growing fast (as most seem to be), talk to their customers about post-sale support. Obtain company reports from impartial sources analyzing long term prospects; look beyond sales and profitability to capital

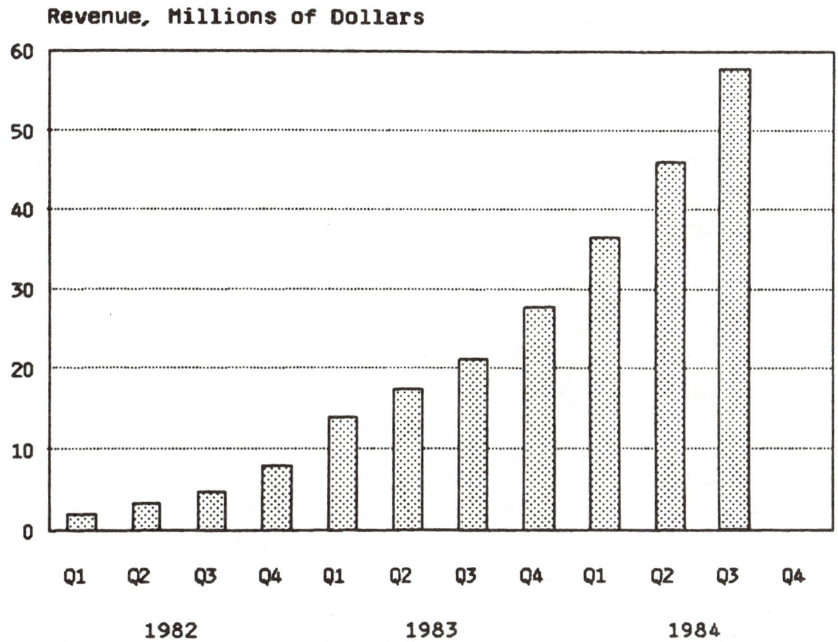

Figure 7.1 Apollo Computer Inc., one of the first companies to produce engineering workstations, has demonstrated through its phenomenal growth record the popularity of dedicated, powerful computers for engineers. (Source: Published reports)

structure, quality and experience of management, investment in research and development: the factors which will determine the company's ability to survive.

Of course, not all workstation vendors are startup companies. Recently, companies like Tektronix, Hewlett-Packard, and Computervision have entered the workstation market in earnest. The presence of companies like these in the workstation marketplace brings a welcome feel of stability to many customers, and will undoubtedly allow them to capture a substantial share of the overall market in the coming years. However, sheer size and experience do not necessarily mean that these companies will be able to cope with the growth to come in engineering workstations any better than the entrepreneurial companies.

7.2 Selection Methodology

Selection of a vendor is obviously a key activity in any capital purchase. Indeed, most companies will expend far more effort in this area than they spend in analysis of their needs or in financial justification. All authorities and texts caution buyers to understand their needs *before* talking to vendors, so the vendors cannot subtly mold potential customers' requirements to the things the particular vendor does well. However, engineering workstations are so new, and their ultimate usefulness so speculative, that potential buyers may be forced to define their needs as they learn what is available.

Engineering workstations enhance the portions of the product development cycle that are least understood by management and by financial people (even by engineers themselves!). The technical qualifications of these products should not be evaluated by people who are not skilled in design engineering, or at least extremely knowledgeable about it. Most companies will form a team to evaluate a new type of product; make sure your workstation team is heavily weighted toward the ultimate users, the engineers. A typical evaluation team will consist of several design engineers (preferably a mixture of senior people and a few "whiz kids"), one or two CAD/CAM specialists (who will probably have responsibility for systems support after the installation), and one person from engineering management (usually the team leader). Specialists in finance and general company management will, of course, be involved, but generally cannot make a meaningful contribution to vendor selection during the technical evaluation phase. These people will become actively involved and make their major contribution toward the end of the evaluation, when investment analysis and overall impact on the business are considered in detail. Beware of allowing financial or management information systems people to choose your workstations; you may end up with the product of the most substantial, conservative vendor, and the poorest technical solution. In this industry, the risky startup companies frequently have the most innovative products.

Vendors of traditional CAD/CAM systems are for the most part substantial companies by now, with solid track records and large customer bases. Selection of a vendor among a group of solid companies can be done with some confidence, since it is easy to get credit ratings, financial analyst's reports, and customer referrals for all the competitors. However, just the opposite is true for many engineering workstation companies. Many of these companies are very young and are owned by entrepreneurs or venture capital firms. It can be nearly impossible to obtain impartial information on the financial health of these companies. In addition, these small companies rarely have a large installed product base from which to obtain referrals; in any event, most customers will have had their equipment such a short time that it will be impossible to learn much of significance about reliability or product support. These, then, are the two areas in which acquisition of workstations differs from acquisition of other types of CAD/CAM equipment: difficulty of measuring the results of using the equipment, and potentially risky vendors. The following sections offer some guidelines to dealing with these problems.

7.3 A Different Way of Conducting Vendor Analysis

The traditional wisdom in conducting a vendor survey holds that you should begin by preparing a detailed specification of your company's needs. This specification is incorporated into a request for proposal document and sent to all vendors who may be qualified, based on judgement and marketing literature. The field is narrowed by eliminating the vendors who do not bid or appear unable to meet the specification. Finally, a benchmark test is conducted to select the "best" vendor.

While the basic methods described above can work for almost any type of product acquisition, most buyers of engineering workstations will find that preparing a meaningful specification is extremely difficult for a product which aims to change the entire way engineers work and think. There are many different approaches to design engineering, and products vary significantly in the ways they attempt to aid the process. This is not to say that you should not define your requirements; what it does mean is that you should only try to define the broad areas of activity to be addressed (e.g., printed circuit board design or finite element analysis of engine parts) and the required interfaces (e.g., to existing systems or software).

Determining which vendors address your problem can be a challenge in itself, given the large number of new companies involved. An excellent way to start is to attend a trade show relevant to your industry, such as Autofact (mechanical engineering) or the IEEE/

ACM Design Automation Conference (electronics). Other good trade shows for finding potential vendors are the National Computer Graphics Association's annual show or the National Computer Conference. Trade journals frequently publish lists of vendors for these products as well. You will discover that, in almost any field of specialty, there will be between 20 and 50 potential vendors.

Since there are so many vendors, many of whom will be entirely new names to you, and each of which seems to have attacked your problem in a different way, the traditional procedure will soon seem impractical. How can a specification be prepared to reflect that some vendors supply only software while some supply turnkey systems? Some utilize mainframes and some use personal computers? Some use general purpose computers and some use high-powered special purpose engines? The best way to sort out the vendors which may solve your problems is to request a complete literature package, screen the vendors by telephone, and make your own judgement.

Many companies require that any vendor contacts be made through the purchasing department. If this is the case at your company, the purchasing department can be used to establish contacts and request literature; however, do not rely on a purchasing department to judge which systems meet your needs. This is a highly technical judgement which may hinge on subjective factors that only come to light when you read what each vendor offers. There will be many surprises and discoveries once you begin to investigate the offerings. Only after you have reviewed a good number of literature packages will you be able to determine what you want in your systems. If you must write a specification, the time to do it is after reviewing what is available, not before.

One of the most exciting things about engineering workstations is that they are usually small and can be brought to your facilities for demonstrations. If at all possible, a live demonstration at your company is preferable to a trip to the vendor's site for the initial contact. This will allow you to invite a larger, more technical audience to see a variety of systems than would be possible if travel were necessary. It also makes it much easier to control the demonstration yourself, to see what you want to see instead of a carefully orchestrated show on the vendor's turf. Only after a successful initial demonstration at your own facility should you spend the time and money to visit vendor sites. This activity should be limited to only a few (say three to five) potential vendors, and you should have firm objectives in mind.

7.4 Vendor Visits

Even though the evaluation team may not have a precise definition of the company's needs, and even though the team may be heavily slanted toward technical people, there should still be a game plan to be followed when visiting each vendor. A constructive vendor visit can be conducted in a single day, and result in everything you need to know to decide whether the vendor should be eliminated.

A vendor visit should consist of four elements: first, the vendor should present an overview of the company, its history, its management, and its philosophy. Second, the vendor should present a product summary, describing the major features and capabilities of its workstation products. Third, there should be a demonstration of the product. Finally, a roundtable discussion should be held in order for the evaluation team to ask questions and clarify what was learned. Beware of vendors who will not do all four sections willingly; they are probably hiding something from you.

One of the best ways to ensure you get all the information you need is to have a list of questions to be answered, revealing the vendor's attitudes, strengths, and weaknesses. Comparing answers to a list of thoughtful questions can eliminate marginal vendors quickly. The questions will not normally be asked directly, but will only serve as a guideline and a means of organizing the information obtained at the vendor's presentations and demonstrations. In the following section, some of the important questions you should consider are described. Each company will have different needs, so be sure you tailor the list to your company's requirements.

7.5 Questions for Vendors

1. How will the system affect the way we currently do business? Will it interface with existing systems, or will it require engineers to learn a whole new way of operating?

2. What sort of improvements in our design cycle will we see? How long will a typical design take compared to current methods? How can we measure the improvement?

3. How much training is required to use the system efficiently? What types of support and training are available from the vendor, both before and after the sale? Is the cost of training and ongoing support included? If so, how long is it provided free? How much does it cost thereafter?

4. Exactly what capabilities are incorporated in the system *now*? When we receive the system, what will we be able to do? (In other words, how much of the demonstration and presentation are "futures"?)

5. What design and network management tools are provided? What capabilities are provided for controlling the data generated on the systems? Can a system management function be built around present capabilities, or must designers take responsibility for their own data archiving and maintenance?

6. What communications capabilities are currently supported? Can the workstations communicate with each other and with other systems? Exactly what kinds of communications tasks can they do?

7. What database access capabilities exist? Can existing designs be brought into the workstation environment? Can the workstations drive existing tools running on different computers?

8. What equipment besides the workstations must be bought in order to use them effectively (e.g., plotters, printers, large discs, interface boards, etc.)? Is this equipment supplied by the workstation vendor? If not, who will guarantee that it all works together correctly? How much does the extra equipment add to the total cost of a workstation?

9. How will the product change in the next six months? The next year? Two years?

10. Who are the vendor's major customers? Who are some of the largest, the most recent, and the most satisfied? (Face it, you will not be told about the unhappy ones!) May we contact the key people at these companies?

11. How do you deal with software licensing on multiple workstations? Must software be purchased for each station in a network, or can it be shared? What sort of price discounts are available for common software on multiple stations?

12. Can I interface third party peripherals and software to a network of your products? How? Who has done it?

13. What parts libraries are available for the system? Who supplies them? Who guarantees that they are correct? How much do they add to the cost of the system?

7.6 Benchmarks

After an initial screening of dozens of vendors, you should have the list narrowed down to a few who seem best suited to the job. This winnowing process is very difficult, and may take many months. It is important to be thorough, but know when to tell a vendor you have seen enough. Do not waste time on a vendor that obviously could not be number one, even if they look like a good second or third choice. You will almost always find two or three finalists, any of which could do the job.

Benchmark tests are intended to select the "best" of a small group of finalists in a competition. However, they rarely accomplish that aim unless the equipment and the task are precisely measurable. The best you can hope for in a benchmark of engineering workstations is to compare a few narrow features (tasks like simulation) between systems for their relative performance. Vendors who supply accelerators and special hardware are likely to apply pressure on prospective buyers to perform benchmarks. However, system selections are rarely decided on totally objective grounds; *do not feel compelled to perform a benchmark*! It is up to you, the buyer, to decide whether to do a benchmark, not to the vendors. If you do a benchmark, plan to spend far more time constructing tests than your initial estimates. Also, prepare ahead of time your criteria for passing or failing, and be prepared to stick to them when the losing vendors protest that you have not evaluated them fairly. In the end, most companies evaluating workstations will choose the vendor they feel the most comfortable with, not the one that performed most functions faster than the others.

7.7 Financial Justification

To most engineers, justification of new equipment is a nuisance. Consequently, it is frequently done poorly, resulting in unnecessary delays in acquiring needed equipment. This section is devoted to a brief overview of two good methods of proving to your company that which you already know: that engineering workstations will be a useful and good investment.

It is beyond the scope of this book to explain the details of investment analysis. However, suffice it to say that the higher in a company's management you must go to justify your acquisition, the more carefully you must prepare your case, and the less relevant the technical details will become. The chairman of the board is not likely to have the time or even the knowledge to read your fifty-page evaluation of the top ten vendors. He is, however, acutely interested in what the financial impact of your proposed investment will be on the company's profitability. A well-prepared investment analysis can be presented in a few pages, and can be backed up by capsule summaries of the vendor selection criteria and the major assumptions used in the analysis.

Before using any financial analysis method, it is crucial that the finance manager and industrial engineering manager at your company be consulted. These individuals will understand how the company views investment proposals, and will usually help prepare your justification. They can provide you with examples of previous successful (and unsuccessful!) proposals to use as models. Furthermore, when the time for presentations and approval signatures comes, these key individuals are likely to be among the people who must approve your proposal. It is of great benefit if they already understand and support your case.

7.7.1 The Productivity Model

The classical approach for justifying equipment which improves productivity is to compare the cost of the equipment with the cost avoided through its use. For instance, suppose your goal is to double the output of a manufacturing operation that is currently performed by machines that require one operator. Your choices are to double the number of machines and people, or possibly to replace the machines with newer types which produce twice the output for the same number of employees. The cost of the first approach is the cost of the machines and the additional personnel; the cost of the second approach is just the cost of the new equipment less the salvage value of replaced older equipment. The difference between the two costs represents a saving in expenses, which is used to determine the financial benefit to the company of investing in the new equipment. Figure 7.2 illustrates such an analysis. The first section (Baseline) states the costs of doing business at present. The next section (Option 1) states the costs of adding more people and equipment. The baseline costs are added, resulting in the total costs of doing business with doubled production using this approach. The next section (Option 2) shows the costs of replacing old equipment with the more productive variety. Again, the baseline cost is added, giving the total operating costs when production is doubled using the second approach.

Finally, the Analysis of Benefits section illustrates the relative consequences of the two approaches to doubling production. Taking the difference between the total costs of the two options, and allowing for taxes on the additional profit, results in the additional after-tax profit resulting from the increased productivity of the newer equipment.

Each company has a different cost structure, and a different way of evaluating savings; in the example given, it is assumed that equipment is straight-line depreciated for 5 years and the tax rate is 50%. These calculations result in the bottom line shown, indicating an increased net profit if investment option 2 is followed. Finally, the present value of the resulting profit stream is calculated, assuming a 12 percent cost of capital. This method of calculating payback is very straightforward, and is familiar to accounting and management people everywhere. Thus, assuming the figures are carefully and conservatively handled, this type of justification can be easy to "sell."

7.7.2 Product Life Cycle Model

The productivity model has been used for years to justify CAD/CAM equipment. For tasks like drafting and production documentation, it is relatively easy to measure productivity increases using traditional time-study methods. However, for the more creative tasks performed on engineering workstations, a certain level of guesswork is inevitably involved. Giving an engineer a computer twice as fast as his previous model may not mean he will design twice as fast; rather, he may do twice as much analysis and produce his design in the same amount of time. Most engineers intui-

Project to Double Production: Analysis of Cost Swings

	1984	1985	1986	1987	1988
Baseline					
Machines (depreciation)	200[a]	200	200	200	200
Operators (burdened salaries)	100	100	100	100	100
Facilities (allocated costs)	100	100	100	100	100
Total operating costs	400	400	400	400	400
Option 1 Double equipment and operations					
Additional machines (same kind)	200	200	200	200	200
Additional operators	100	100	100	100	100
Additional facilities	100	100	100	100	100
Additional operation costs	400	400	400	400	400
Total operating cost, Option 1	800	800	800	800	800
Option 2 Replace machines with new ones which are twice as productive					
All new machines (2X productivity)	400	400	400	400	400
Additional operators (none)	0	0	0	0	0

Additional facilities (upgrade)	100	0	0	0	0
Operator training	50	0	0	0	0
Additional operating costs	550	400	400	400	400
Baseline operating cost	400	400	400	400	400
Less depreciation for old machines	(200)	(200)	(200)	(200)	(200)
Total operating cost, Option 2	750	600	600	600	600
Analysis of Benefits					
Total cost of Option 1	800	800	800	800	800
Less total cost of Option 2	(750)	(600)	(600)	(600)	(600)
Less 50% tax on additional profit	(25)	(100)	(100)	(100)	(100)
Net additional after-tax profit	25	100	100	100	100
Present value of profit stream	$293K				

[a]All figures in thousands of dollars

Figure 7.2 Many companies will be able to use a classical productivity model to justify engineering workstations. The increased output of the systems permits more work to be done without additional human resources.

tively know that the resulting design will be better, but how can that be translated into dollars and cents?

One method frequently used to quantify better design is to assume that such designs will require less rework, or fewer prototype passes, due to more thorough simulation, better documentation, etc. Thus, even though the design may take just as long, the probability of doing it right the first time is increased. If, as is common in some industries, a "recycle" is planned into the development schedule, getting it right the first time translates into a shorter overall development cycle. This, in turn, can be interpreted in terms of more work output per engineer, or shorter time to design.

If it is desirable to interpret the data as more work output per engineer, the productivity model may be used. Since it is unclear at the outset just how often a recycle will be avoided, it is better to work the problem backward: find what percent improvement in recycles will result in breaking even on the investment. Then decide, based on judgement, whether it is reasonable to assume you will do better than this percentage. This type of analysis is easily implemented using a goal-seeking financial modeling system, such as is found in many "spreadsheet" software packages.

If the data are interpreted as shorter time to complete a design, the product life cycle model can be used effectively. The theory underlying this justification is that a particular product, all other factors being equal, will become obsolete around the same date regardless of its date of market entry. If this assumption is true (which it certainly is in many industries), then the overall revenue realized over a product's life cycle is greater if it is introduced earlier. A strategy for justification of engineering workstations, then, is to evaluate the reduction in design time to be realized through their use, and trade off their cost against the additional revenue realized from earlier market entry. Industries in which this principle is valid include most high technology commodity markets, such as games, toys, semiconductors, entertainment electronics, cameras, watches, and the like. The principle should be used with caution in capital equipment, transportation, and other industries where product obsolescence and technological innovation do not happen so quickly.

In many industries, a product's life cycle follows the well-known pattern seen in marketing texts: rapid rise in volume, followed by a maturity period, followed by a declining period. The semiconductor industry provides a textbook example, although the principles apply for many other industries. For commodity semiconductor devices like memory and microprocessor parts, this cycle may last from three to seven years. Accompanying the market maturity pattern is a pricing pattern characterized by rapid decline from an initial price determined mostly by the first vendor to supply a part. Figure 7.3 illustrates these two trends in general form. Figure 7.4 illustrates the resulting sales revenue from the multiplication of the two curves, illustrating that the early years of a product's life generate most of its revenue.

Now, suppose that the hypothetical product is introduced six months early, owing to greater design productivity. What actually appears to happen is that the left edge of the sales volume curve shifts to the left, and the "tail" of the curve at the right remains fixed. The area under the new portion of the curve on the left represents net revenue which would never have been obtained had the productivity gains not been realized.

Justification of your engineering workstations using these facts requires only some fairly straightforward calculations, using numbers readily available from strategic marketing and finance departments in most companies. Use of a spreadsheet program can ease this process immensely, but it can also be done by hand. Figure 7.5 illustrates the results of such an exercise.

First, a model of the life cycle is built (Original Plan), resulting in a revenue figure for each year (or month, or whatever) over the life cycle. Then, a second model is built (Early Introduction Plan), allowing for the earlier market entry, and producing a new revenue curve. The difference between the results of the two models produces the Improved Cash Flow line, the contribution expected from each new product introduced early. The following section of the illustration summarizes the effects of all new products that will be affected by early introduction, and results in a total cash flow. Assuming a gross margin of 50 percent and a tax rate of 50 percent, the after-tax profit improvement resulting from the project is calculated. The cost of the equipment is subtracted from the gross revenue, giving the net cash flow after

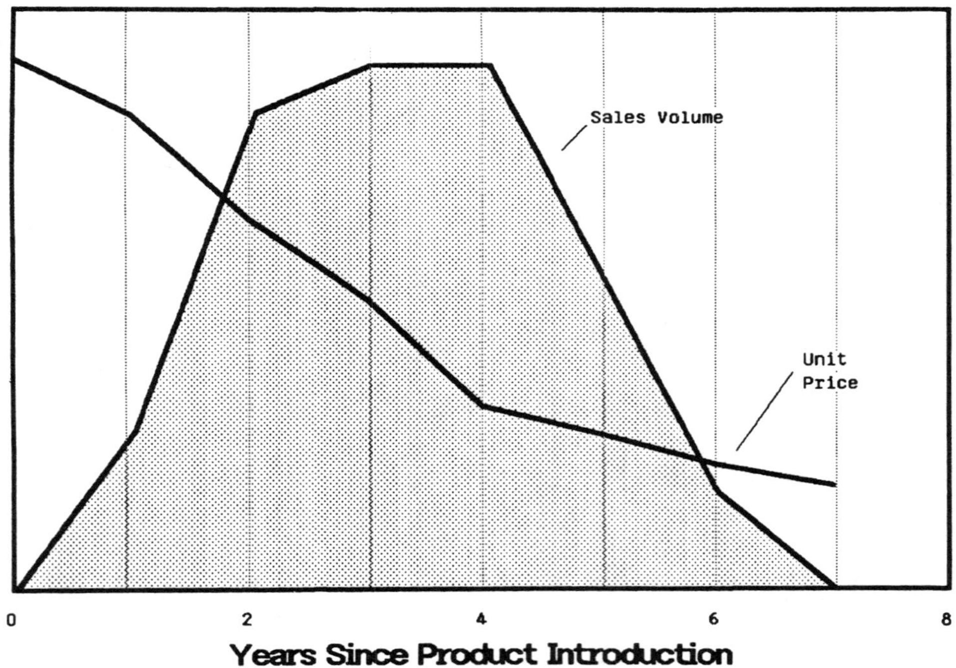

Figure 7.3 The commodity semiconductor industry is characterized by a short life cycle and rapidly declining prices. (Based on author's estimates.)

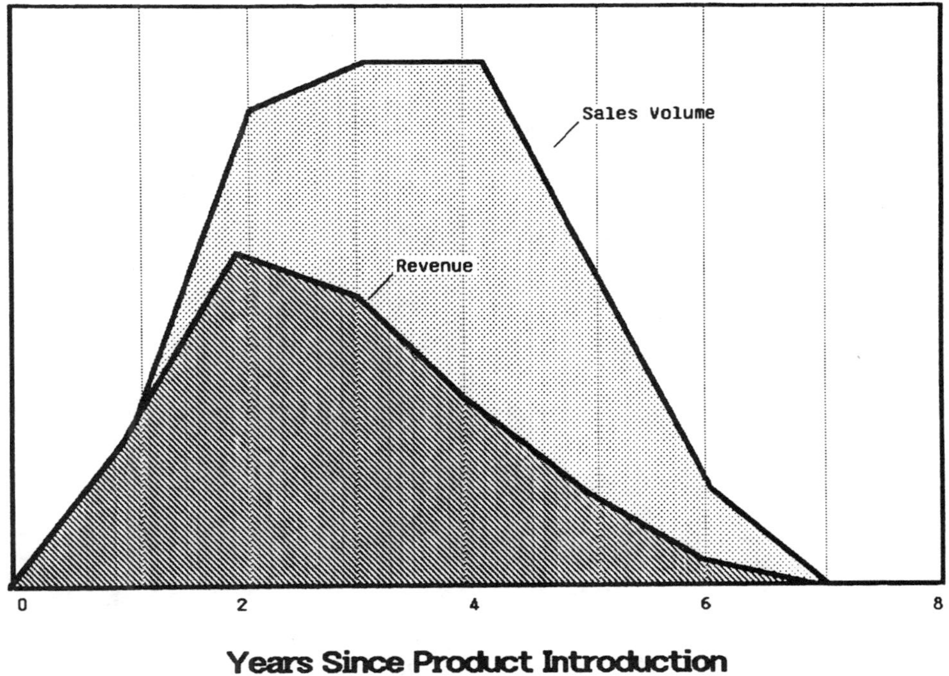

Figure 7.4 Revenue over the life of a commodity semiconductor component is heavily weighted toward the early years of the product's life. (Based on author's estimates.)

Project to Introduce Products Earlier: Profit Analysis

Return on Investment Calculation	1984	1985	1986	1987	1988	1989	1990	1991
Original Plan (per design)								
Volume (K units)		200	1,000	2,000	2,000	1,600	600	200
Average selling price ($)		$1.00	$.75	$.50	$.50	$.37	$.30	$.25
Revenue (per design, $K)		200	750	1,000	1,000	600	180	50
Early Introduction Plan (per design)								
Volume (K units)	15	376	1,692	2,000	2,000	1,600	600	200
Average selling price ($)	$1.25	$1.00	$.75	$.50	$.50	$.37	$.30	$.25
Revenue (per design, $K)	19	376	1,269	1,000	1,000	600	180	50
Improved Cash Flow (per design)	19	176	519	0	0	0	0	0
New Designs Affected (per year)	2	5	10	15	20	20	20	20
Total cash flow (see Exhibit A)	38	980	7,153	10,730	14,307	14,307	14,307	14,307
Resulting gross margin improvement	19	490	3,576	5,365	7,153	7,153	7,153	7,153
Resulting after-tax income improvement	9	245	1,788	2,682	3,576	3,576	3,576	3,576
Equipment Investment ($K)	900	1,250	400	250	250	200	150	100
Less Investment Tax Credit	90	125	40	25	25	20	15	10
Net Investment ($K)	810	1,125	360	225	225	180	135	10
Annual Net After-Tax Cash Flow ($K)	(800)	(879)	1,428	2,457	3,351	3,396	3,441	3,486
Present Value of Profit Stream ($K)	$7,750K							

Exhibit A Cash Flow Contribution by Year from New Designs:

Year 0	19	176	519	0	0	0	0	0
Year 1	0	19	176	519	0	0	0	0
Year 2	0	0	19	176	519	0	0	0
Year 3	0	0	0	19	176	519	0	0
Year 4	0	0	0	0	19	176	519	0
Year 5	0	0	0	0	0	19	176	519
Year 6	0	0	0	0	0	0	19	176
Year 7	0	0	0	0	0	0	0	19
Total contribution ($K)	38	980	7,153	10,730	14,307	14,307	14,307	14,307

Exhibit B Assumptions:

Weeks early (result of investment)	4	10	20	20	20	20	20	20
Tax bracket (%)	.5							
Gross profit margin (%)	.5							
Investment tax credit (%)	.1							
Volume factor for model		.1	.5	1	1	.8	.3	.1
Price factor for model	2.5	2	1.5	1	1	.75	.6	.5
Annual sales volume at maturity (million units)	2.0							
Unit price at maturity ($)	$.50	$.70	$.85	$1.00	$1.00	$1.00	$1.00	$1.00
Cost of capital	12%							

Figure 7.5 A spreadsheet program simplifies the task of analyzing the payback of engineering workstations using the life cycle model. (Source: author's estimates, for illustrations only.)

taxes, the additional profit resulting from the use of the equipment. Some companies require computation of the present value of the profit stream, as illustrated in the lower left of the spreadsheet; here a 12 percent cost of capital was assumed.

Since each company does its financial analysis a little differently, there is no one program or method that is always right. Use the illustration in Figure 7.5 simply as a guide to the overall methodology. Naturally, with a spreadsheet program and a carefully constructed model, it is easy to try different productivity levels, different price curves, and so on. Generally, your first try will result in numbers so favorable that you will be forced to make all assumptions more pessimistic in order that they will be believed!

8

TRENDS IN THE ENGINEERING WORKSTATION INDUSTRY

Potential engineering workstation buyers should ask questions not only about their current needs, but about the future: what new features and capabilities can be expected, and when? Which vendors will dominate the marketplace? How much will future systems cost compared to what they cost now? Is this the right time to buy, or can I afford to wait? Such questions, of course, have no simple or easy answers; however, understanding some facts about the size of the market, its composition, and some of the economics of workstation use can help provide some useful insight.

8.1 Industry Size and Growth

The overall CAD/CAM market is forecast [1] to grow from $2 billion in 1983 to $11 billion by 1988, a compound annual growth rate of 40 percent. This forecast alone illustrates the huge commitment industry is making to automation in both design and manufacturing; even more revealing to potential buyers of engineering workstations is the breakdown of the types of equipment that will make up the CAD/CAM market. In 1983, standalone workstations and systems based on personal computers made up only about 15 percent of all CAD/CAM stations sold; the vast majority of sales were of host-dependent systems. By 1988, over 40 percent of sales will be attributed to standalone and personal computer based workstations.

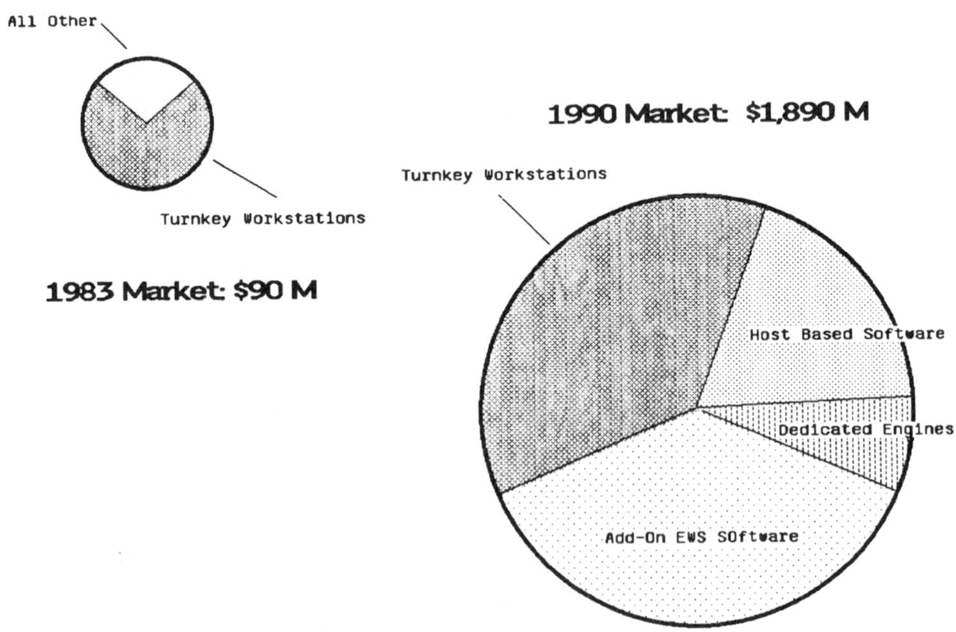

Figure 8.1 Electronics industry spending for CAE workstations is forecast to grow explosively for the remainder of the current decade. (Source: *EDN Magazine*)

Within the huge total CAD/CAM market, there are several major submarkets. Electronics CAD/CAM, while remaining a relatively small portion of overall CAD/CAM sales, indicates what may happen to other segments when economical, powerful engineering workstations become available. One estimate [2] indicates that electronics CAE equipment and software sales will rise from only $90 million in 1983 to nearly $2 billion by 1990. Of this total, over 80 percent is expected to be for standalone workstations, dedicated engines, and add-on workstation software, with less than 20 percent going for host-based software. Figure 8.1 illustrates this staggering commitment to engineering workstations.

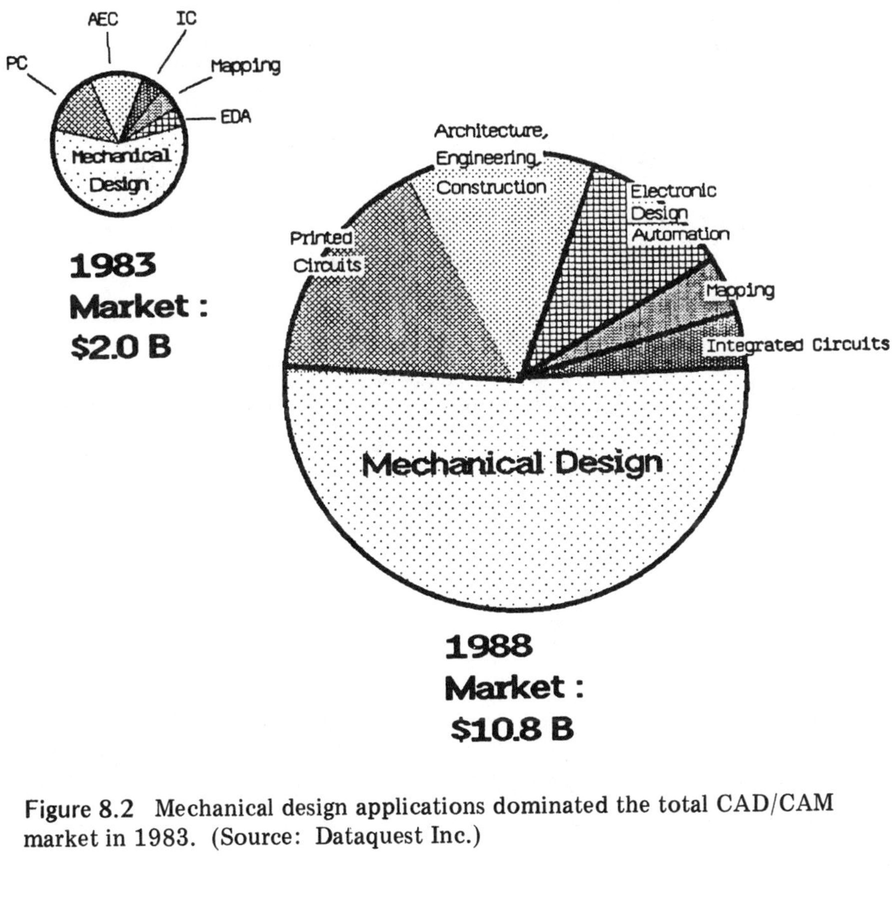

Figure 8.2 Mechanical design applications dominated the total CAD/CAM market in 1983. (Source: Dataquest Inc.)

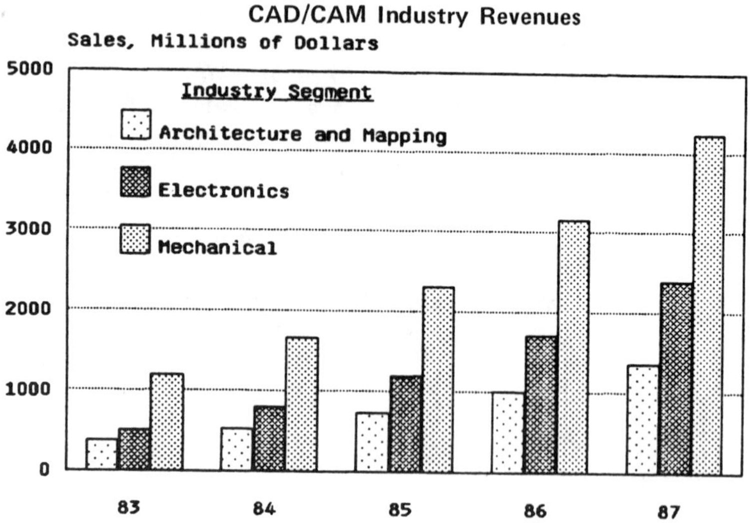

Figure 8.3 By 1990, mechanical designers will endorse engineering workstations as enthusiastically as electronics engineers have done in the early 1980s. (Source: Dataquest Inc.)

The overall CAD/CAM marketplace is dominated by mechanical design applications (Figure 8.2). Most authorities expect this to continue indefinitely. In 1983, mechanical design provided 58 percent of CAD/CAM industry revenue, while electrical and electronics together provided only 23 percent. Figure 8.3 illustrates one estimate [1] of the breakdown of the total market. If mechanical engineers embrace engineering workstations as enthusiastically as electronics engineers have done, there is potentially a $5 billion market, representing nearly 100,000 mechanical engineering workstations, in the early 1990s.

8.2 Technology Trends

Mechanical design will continue to represent more than half the total CAD/CAM market for the foreseeable future, and engineering workstations will be a large part of this market. Today, limitations in computing power prevent widespread use of workstations in this industry, but the situation is changing rapidly. A recent report on manufacturing automation by the congressional Office of Technology Assessment predicts that systems powerful enough to support "onboard" finite element analysis and realtime generation of shaded solid images will be available in the 1987 to 1990 timeframe, and in widespread, routine use by the mid-1990s. Much of this progress will be the result of the development of application-specific hardware for image generation and finite element analysis, developments of the electronics industry made possible partly by their adoption of engineering workstations!

In the late 1990s, comprehensive engineering workstations will incorporate a "window into the factory," a firm link between design decisions and manufacturing consequences. Designers using such systems will be able to rapidly try different design approaches, and to understand their impact on the entire product. Systems will eventually be powerful enough to permit extensive interactive testing, simulation, and refinement of designs in a wide variety of applications. Today, such capabilities exist only in a few highly specialized applications, and are very product dependent (e.g., semiconductors, automotive subsystems). Comprehensive systems providing full design-to-factory integration in all industries are not expected to be widely and easily available until at least the year 2000. Commercially available electronic design systems may reach this level of capability by 1987, at least for companies having highly automated manufacturing processes. Mechanical design systems are expected to lag by as much as a decade or even more, owing to the vastly more complex and varied manufacturing processes with which the systems must deal.

A significant trend in the workstation industry is the gradual adoption of artificial intelligence (AI) technology. AI is a term which conjures up many different visions, but in this context, it usually refers to expert systems, which use a knowledge base to assist designers with the subjective portions of their job. Over the next five years AI will begin to appear as design-assistance built into the workstation.

An expert system typically contains a body of facts and rules of thumb about a given subject (a knowledge base) and a software system for making inferences. In principle, such systems can be configured so as to learn from experience, storing in the knowledge base information on the user's action, thus encapsulating many of the personal preferences and methods of a given designer. For example, in the conceptual design of an office building by an architect, an expert system might lead the designer through a series of questions on the number of occupants, the different departments or organizations to be housed, their relationships, and so on. It might then be able to suggest an optimum number of floors based on the parameters and the site restrictions. In a similar way, the designers may progress from generalities to details with the system acting as partner rather than slave.

Artificial intelligence is not as far-fetched as it sounds; successful expert systems in fields as diverse as petroleum exploration, medical diagnosis, and linguistics are in routine use today. Design engineering is an ideal application for AI technology, dealing as it does with a large body of fact, many rules, and judgement factors. Many experts in the field feel that engineering workstations will routinely exploit AI principles by the end of the 1980s.

A major trend in the software for engineering workstations is the concept of correctness by design. If the tools used by a designer know about the rules and constraints on his implementation, they can, in principle, prevent many, if not all, mistakes. While an expert system may be the ultimate solution, there are many tools available and being developed now that address the problem. In electronics, for instance, workstations will create an environment in which design and testability rules, initial timing, and functionality are incrementally checked as schematic and parametric data are entered. Many such capabilities are already appearing, and it is reasonable to expect their widespread use by the latter part of the decade.

One of the major benefits of engineering workstations in the future will be the ease with which a design may be refined and options may be explored. One vendor (CAE Systems Inc.) offers DCALC™, the "designer's calculator," a capability for engineers analogous to a financial spreadsheet program. The designer builds a table of information based on relationships among design data, and DCALC then shows the results of a variety of what-if experiments without actual alteration of the design. A logical extension of this capability, given enough computing power, is a free-form optimization capability which would allow designers to find better designs through improvement of parameters; the increased computational power of future workstations will make such capabilities feasible.

8.3 Product Life Cycle: A Crisis in Productivity

While all industries are pressed to increase their productivity today, none provides so graphic an illustration of the crisis as the electronics industry. Perhaps no other industry in history has seen such rapid technological evolution in its products, nor such rapid obsolescence. While the pace of product obsolescence quickens, the complexity of the products themselves grows exponentially. Figure 8.4 illustrates the problem faced by this industry: if product development time cannot be reduced drastically and continuously, many companies will become noncompetitive and are sure to fail.

Engineering workstations are not the only answer to the problem, but it is certain that they can be a major contributing factor to the resolution of the dilemma described above. Mentor Graphics Corporation, a leading manufacturer of electronic workstations, conducted a study indicating that in 1980, electronics firms spent an average of only $12,000 per employee for capital equipment, while the auto industry spent $20,000, the chemical industry $46,000, and the petroleum industry $210,000. Mentor estimates that, in order to remain competitive, electronics companies will be spending $150,000 per employee by 1990.

Many other industries will soon feel this sort of pressure. Industries like automobiles, machine tools, and many consumer goods manufacturers are currently beseiged by competitive pressure from countries enjoying low labor costs. These industries must employ the most advanced CAD/CAM techniques available in order to remain competitive; engineering workstations are sure to be a large part of their investment in productivity equipment in the late 1980s.

A recent survey by EDN Magazine [2] indicated that the average engineer or engineering manager in the electronics industry would willingly spend $16,700 per engineer for CAE tools, or $25,200 if integrated circuit layout software were included. At the time of the survey, the average CAE workstation for electronic design cost about $50,000. The survey indicated further that, on the average, 3.5 engineers were sharing a workstation in 1984 in companies that had al-

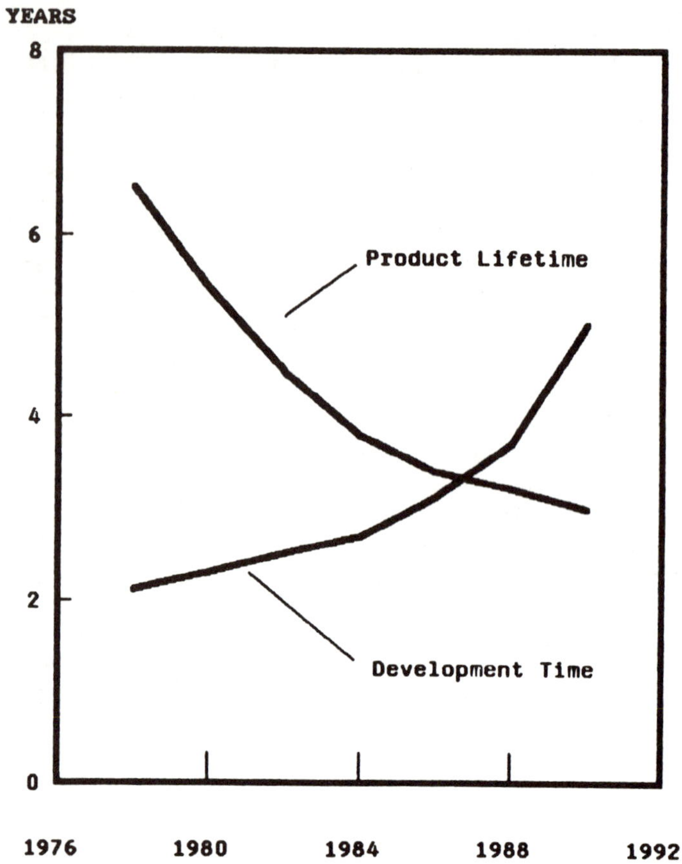

Figure 8.4 The time it takes to develop a new high-technology product will soon exceed the useful life of similar products. Companies will utilize the best available tools to increase engineering productivity as a matter of competitive necessity. (Source: Masscomp Inc.)

ready bought the equipment. Although it is risky to draw conclusions from this type of information, it is safe to guess that most companies still feel that engineering workstations are too expensive to put on every engineer's desk.

However, the cost of engineering talent is increasing very rapidly, and many companies will discover that the increased productivity of scarce engineers is well worth the capital expenditure. The American Electronics Association predicts that by 1985, the demand for new electronics engineers will exceed 50,000 people, yet American universities will graduate only about 15,000. These facts assure that the cost of engineers will continue to rise rapidly for the foreseeable future.

The economic payback illustrations in Chapter 6 clearly indicate that workstations, even at the $50,000 level, are justifiable today; with the burdened cost of an engineer headed for the $100,000 level in the middle 1980s, and a growing crisis in product development cycles, a workstation which doubles the engineer's productivity may soon appear to be a necessity, not a nicety.

References

1. Focus Conference, Dataquest Inc., "Workstations: The Evolution of an Industry," September 21, 1984, Boston, MA.
2. *Implications of Computer-Aided Engineering: An EDN Report.* Cahners Publishing, Boston, MA, 1984.

APPENDIX: VENDORS OF ENGINEERING WORKSTATIONS

The following list was compiled in early 1985. As explained elsewhere in this book, vendors of engineering workstations appear and disappear daily; thus a list like this one should not be regarded as complete, and may be inaccurate. Readers are urged to consult any of the periodicals and newsletters in their fields for more current information. Some of the better specialized newsletters known to the author are listed at the end of this appendix.

The author apologizes in advance to any vendor who has been omitted, or whose address is stated inaccurately.

A.1 General Purpose Systems

The following vendors produce computers used by workstation vendors. In most cases, these consist strictly of hardware and system level software, and are not integrated systems. However, in many cases, software is available from third parties that may allow sophisticated users to configure their own workstation environment.

Apollo Computer, Inc.
330 Billerica Road
Chelmsford, MA 01824

Celerity Computing
160 Speen Street
Framingham, MA 01701

Data General Corporation
4400 Computer Drive
Westboro, MA 01580

Hewlett-Packard Company
1820 Embarcadero Road
Palo Alto, CA 94303

Ridge Computers
2451 Mission College Blvd.
Santa Clara, CA 95054

Saber Technology
2381 Bering Drive
San Jose, CA 95131

Sequent Computer Systems Inc.
14360 Science Park Drive
Portland, OR 97229

Sun Microsystems
2550 Garcia Avenue
Mountain View, CA 94043

A.2 Systems for Personal Computers

The following vendors supply either software the user must install, or complete systems including the computer. Unless otherwise noted, all use the IBM PC or compatible systems.

Autodesk Inc.
2658 Bridgeway
Sausalito, CA 94965

The CAD Group
P.O. Box 383
830 Bay Avenue, Suite G
Capitola, CA 95010

CADLINC
700 Nicholas Blvd.
Elk Grove Village, IL 60007

The CADWARE Group, Ltd.
869 Whalley Avenue
New Haven, CT 06515

Case Technology, Inc.
444 Castro Street
Mountain View, CA 94041

Chancellor Computer
1101 San Antonio Road
Mountain View, CA 94043

Control Data Corporation
Information Systems Design
2500 Mission College Blvd.
Santa Clara, CA 95054

Datagraphic Systems
9101 General Drive
Plymouth, MI 48170

E/Z-CAD Inc.
5589 Starcrest Drive
San Jose, CA 95123

FutureNet Corporation
21018 Osborne Street
Canoga Park, CA 91304

MacNeal Schwendler Inc.
815 Colorado Blvd.
Los Angeles, CA 90041

Mega CADD, Inc.
2001 Sixth Avenue
Seattle, WA 98121

MicroSim Corporation
14101 Yorba Street, Suite 202
Tustin, CA 92680

Mindset
617 N. Mary
Sunnyvale, CA 94086

Personal CAD
981 University Avenue
Los Gatos, CA 95030

RACAL-REDAC Inc.
Lyberty Way
Westford, MA 01886

The Rand Group, Inc.
17430 Campbell Road, Suite 114
Dallas, TX 75252

ROBO Systems (Apple II)
Chessell-Robocom Corporation
111 Pheasant Run
Newtown, PA 18940

Vivid Software Corporation
900 Lafayette Street, Suite 604
Santa Clara, CA 95050

Wintek Corp.
1801 South Street
Lafayette, IN 47904

A.3 Full-Function Systems

The following vendors supply fully integrated hardware and software for one or more spplication areas.

Analog Design Tools, Inc.
800 Menlo Avenue, Suite 217
Menlo Park, CA 94025

Auto-Trol Corporation
12500 North Washington Street
Denver, CO 80233

Bausch & Lomb
Box 14547
Austin, TX

BruningCAD
6111 East Skelly Drive
Tulsa, OK 74135

Cademic Inc.
P.O. Box 4698
Scottsdale, AZ 85261

Cadnetix Corporation
5797 Central Avenue
Boulder, CO 80301

CADTEC Corporation
2355 Old Oakland Road
San Jose, CA 95131

CAE Systems Inc.
1333 Bourdeaux Drive
Sunnyvale, CA 94089

Calcomp Inc.
2411 West La Palma Avenue
Anaheim, CA 92801

Calma Company
2901 Tasman Drive
Santa Clara, CA 95050

Cascade Graphics Development
70 Walnut Street
Wellesley, MA 02181

Chromatics Inc., CAECO Division
P1160 S. State St.
Suite 280
Orem, UT 84057

Computervision Corporation
201 Burlington Road
Bedford, MA 01730

Control Data Corporation
2500 Mission College Blvd.
Santa Clara, CA 95054

Daisy Systems Corporation
139 Kifer Court
Sunnyvale, CA 94086

Holguin & Associates
5822 Cromo Drive
El Paso, TX 79912

Intergraph
One Madison Industrial Park
Huntsville, AL 35807

McDonnell Douglas Automation Company
Box 516
St Louis, MO 63166

Mentor Graphics Corporation
8500 S. W. Creekside Place
Beaverton, OR 97005-7191

Metheus-CV Corporation
P.O. Box 1049
Hillsboro, OR 97123

Phoenix Data Systems
80 Wold Road, 6th Floor
Albany, NY 12205

RACAL-REDAC Inc.
Lyberty Way
Westford, MA 01886

Silvar-Lisco
1080 Marsh Road
Menlo Park, CA 94025

Silma, Inc.
1800 Embarcadero Road
Palo Alto, CA 94303

Skok Systems
222 Third Street
Cambridge, MA 02142

TriCAD Inc.
1655 McCarthy Blvd.
Milpitas, CA 95035

Valid Logic Systems Inc.
1395 Charleston Road
Mountain View, CA 94043

Versatec, Inc.
2710 Walsh Ave.
Santa Clara, CA 95051

VIA Systems
76 Treble Cove Road
North Billerica, MA 01862

A.4 Software Vendors

The following vendors sell only software, and it is assumed that the user will supply appropriate hardware. In many cases, vendors will assist the user in selecting and configuring appropriate hardware. Several of the major packages will run on a variety of the systems listed above as general purpose systems.

ARCAD
445 South Figueroa Street
Los Angeles, CA 90071

Cericor, Inc.
716 E. 4500 South
Salt Lake City, UT 84107

General Electric
120 Erie Blvd., Suite 100
Schenectady, NY 12345

MacNeal Schwendler Inc.
815 Colorado Blvd.
Los Angeles, CA 90041

Manufacturing and Consulting Services, Inc.
9500 Toledo Way
Irvine, CA 92718

Tasvir Corporation
100 Bush Street, 11th Floor
San Francisco, CA 94104

UniCAD Inc.
1695 38th Street
Boulder, CO 80301

A.5 Further Sources

The following publications are recommended to readers as good places to find up-to-date information on vendors and systems.

CAD/CIM Alert (Newsletter)
CAE Workstation Alert (Newsletter)
The PC Software Systems Directory for Computer-Aided Engineering (Book)

The above are available from:

Management Roundtable Inc.
824 Boylston Street
Chestnut Hill, MA 02167

Additional publications:

Computer Aided Design Report (Newsletter)
841 Turquoise Street, Suites D&E
San Diego, CA 92109

The Anderson Report (Newsletter)
Anderson Publishing Company
4525 E. Industrial Street, Suite 4L
Simi Valley, CA 93063

Personal Engineering and Instrumentation News (Newsletter)
Personal Engineering Communications
Box 983, Back Bay Annex
Boston, MA 02117-0983

GLOSSARY

The engineering workstation industry, like most other highly technical fields, employs a great deal of jargon. Since many readers of this work will not be familiar with the language of computers or the workstation field, the following glossary is provided. It is by no means complete, but it should cover most of the new terminology introduced in the text.

Accelerator A specialized hardware system whose role is to make one particular function of an engineering workstation run much faster than software.

ACM The Association for Computing Machinery, a professional society.

AEC A CAD/CAM industry term meaning architecture, engineering, and construction.

AI Abbreviation for artificial intelligence.

Algorithm A formalized expression of the method used to solve a problem, frequently used in discussions of software.

ANSI The American National Standards Institute.

Artificial Intelligence A branch of computer science whose goal is to develop systems which mimic the methods and behavior of humans.

ASCE American Society of Civil Engineers, a professional society.

Asynchronous A term used in discussion of data communications, referring to the fact that information is transmitted and received without regard for fixed time intervals. Asynchronous communications are usually significantly slower than synchronous.

Behavioral Simulation Behavioral simulators model the way systems behave, rather than the way they are built.

Benchmark A carefully constructed series of tests, intended to objectively measure the relative performance of several systems.

Bit The smallest unit of information processed by a computer. One bit represents the result of a single yes-or-no decision.

Breadboard A term used by electronic engineers to describe a crude prototype of a circuit being developed.

Byte A unit of data, usually representing one letter or character, in a computer's memory.

CAD Computer-Aided Design.

CAE Computer-Aided Engineering.

CAM Computer-Aided Manufacturing.

CIM Computer-Integrated Manufacturing.

Database A set of files and data which completely describes a system or project.

DRC Design rule checking, the process of auditing a database for compliance with an established set of rules for correct design.

EDIF Electronic Design Interchange Format.

ERC Electrical rules checking, the process of auditing a database for compliance with an established set of rules for correct circuit performance.

Ethernet™ A trademark of Xerox Corporation. A popular network system which is used by many engineering workstation manufacturers.

Glue A colloquial term used by engineers to describe software or hardware which "holds subsystems together."

Icon A small pictorial image used to represent functions or programs on the screen of a workstation. For instance, a small picture of a trash can is a frequently-used icon for deletion of a file.

IEEE The Institute of Electrical and Electronics Engineers, a professional society.

IGES Initial Graphics Exchange Specification.

Integrated Circuit A tiny electronic component which contains hundreds or thousands of permanently connected transistors and other devices. Also known in the popular press as "chips" or "microchips."

Kinematics A term, used in mechanical design CAD, for software which analyzes and displays the relative motion of parts in assemblies.

Mainframe A loose term used for large computers. While the technical distinction between a mainframe and other types of computers is beyond the scope of this book, it is useful to think of mainframes as computers costing over a million dollars.

Megabyte A million bytes; a term used to describe how much memory a computer or peripheral has.

Minicomputer A loose term used for medium-sized computers. As with mainframes, the technical definition of a minicomputer is beyond the scope of this book. Although imprecise, it is useful to think of a minicomputer as a system which supports two to eight simultaneous users, and costs less than $200,000.

Netlist A textual description of the connections among a set of components, used to translate a design into a form useful to a simulator.

Network A system used to connect multiple computers together so that they may communicate and exchange information.

Numerical Control A class of semiautomatic manufacturing machines which are programmed by tapes containing precise instructions detailing movements and tool speeds. Frequently abbreviated N/C or NC in the literature.

Online Refers to operations which are performed directly by the user while connected to a computer, as opposed to operations which are submitted for later processing.

Operating System The software which controls the low level functions of a computer.

Oscilloscope A laboratory instrument used by electronics engineers to examine graphically the behavior of circuits over very short time intervals.

Peripheral Device A device attached to a computer for purposes other than computing. Memory devices, printers, magnetic tape drives, and plotters are common examples.

Physical Layout The process of designing the geometric representation of a printed circuit or integrated circuit.

Pixel One resolvable point on the screen of a terminal or workstation; an abbreviation derived from "picture element."

Plotter A peripheral device which draws images on paper or other drafting media.

Postprocessing The manipulation of design data after the actual design is complete, usually for transfer to a different system or device.

Properties Parameters and attributes of a design which are associated with components, but not usually visible on drawings.

Schematic A diagram using symbols to represent the elements of a design.

Semiconductor Refers to a class of electronic devices, usually integrated circuits, made from special materials which are neither conductors nor insulators.

Silicon The most common semiconductor material used to make integrated circuits.

Simulator A software system used to model the performance of a physical system. Simulators are used during the design phase of many types of products to avoid the expense and time of building prototypes.

Spreadsheet A class of computer programs which deals with rows and columns of numbers, commonly used for financial modeling.

Standalone Refers to a system or computer which does not need to be connected to any other system in order to perform its job.

Supercomputer A very large computer system, typically costing millions of dollars, and capable of processing very large problems in reasonable times.

Superminicomputer An imprecise term for a large minicomputer. For purposes of understanding, systems supporting ten to fifty simultaneous users and costing between $200,000 and $500,000 fall into this category.

Synchronous A term used in data communications to describe systems in which data transmission and reception are coordinated by a time reference or clock. Synchronous communications systems are usually significantly faster and have a lower error rate than asynchronous systems.

Transistor The fundamental element of an integrated circuit, used as a switch or amplifier.

Turnkey System A system which is delivered to the customer in a form that can be used immediately for a specific type of job. Manufacturers of such systems provide all the software, hardware, training, and support needed by the customer.

UNIX™ A trademark of American Telephone and Telegraph Corporation. A popular operating system used on many engineering workstations.

VAX™ A trademark of Digital Equipment Corporation. A family of superminicomputers which are in very widespread use among engineering companies.

VLSI Very large scale integration, an industry term for integrated circuits containing more than about 10,000 circuit elements.

Window A small portion of the screen on a workstation which represents a single view of a drawing or text. Systems using windows allow several simultaneous views of a design, providing great flexibility of usage.

Workstation A system typically consisting of a small computer, a set of application software, and a graphics display, intended for the use of an individual. In the text, products representing many variations on this definition are discussed.

INDEX